Sensors

Topical Conference at the 2012 AIChE Annual Meeting

Pittsburgh, Pennsylvania, USA
28 October - 2 November 2012

ISBN: 978-1-62276-746-5

Printed from e-media with permission by:

Curran Associates, Inc.
57 Morehouse Lane
Red Hook, NY 12571

Printed by Curran Associates, Inc. (2013)

For permission requests, please contact AIChE
at the address below.

AIChE
3 Park Avenue
New York, NY 10016-5991

Phone: (203) 702-7660
Fax: (203) 775-5177

www.aiche.org

Additional copies of this publication are available from:

Curran Associates, Inc.
57 Morehouse Lane
Red Hook, NY 12571 USA
Phone: 845-758-0400
Fax: 845-758-2634
Email: curran@proceedings.com
Web: www.proceedings.com

TABLE OF CONTENTS

(606a) Optical Enzymatic Biosensor for Sensitive Detection of 1,2,3-Trichloropropane

Brian Heinze and **Kenneth F. Reardon**, Chemical and Biological Engineering, Colorado State University, Fort Collins, CO

The emerging water contaminant 1,2,3-trichloropropane (TCP; also known as allyl trichloride, glycerol trichlorohydrin, and trichlorohydrin) is used as an industrial solvent, as a cleaning and degreasing agent, and in the production of pesticides. TCP is a known toxin and has been determined by the State of California to cause cancer. The State of Hawaii has established a state MCL of 0.6 μg/L, and the California Department of Public Health has established a notification level of 0.005 μg/L. A significant need exists for devices that can measure the concentration of TCP and other chemical contaminants in ground and surface waters, and in the influent and effluent streams of water remediation and treatment facilities. These measurements must be rapid, accurate, and inexpensive. Although laboratory-based methods relying on chromatography can yield very accurate measurements, they are also complex, time consuming, require sample pre-treatment, and expensive. Furthermore, they are not readily adapted for in-situ measurements. In contrast, biosensors can provide simple, rapid, in situ, inexpensive measurements that do not require pretreatment of the sample.

We have developed optical enzymatic biosensors for the sensitive measurement of TCP concentrations in aqueous solutions. In our biosensors, a two-layer detection element – one layer containing the detection enzyme affixed to another fluorophor-containing layer – immobilized on the distal end of an optical fiber, which is connected to a light source/detection unit. The change in fluorescence depends on the contaminant concentration, and these sensors provide quantitative output over a range of analyte concentration. Performance metrics including the limit of detection, impact of environmental factors such as pH and temperature, and lifetime have been determined.

(606b) Smartphone Integrable High Density Nano Gas Sensor Array for Pollutant Detection

Lauren Brook, Heng Su, Miluo Zhang, Albert Chen and **N. V. Myung**, Chemical and Environmental Engineering, University of California, Riverside, Riverside, CA

Conventional methods for detection of pollutants are often expensive and time consuming. Data collections for air pollutants are generally confined to air monitoring stations, which are then assumed to be representative of the study region. Laboratory costs for these tests exceed $100/sample and only provide a single snapshot in time. This lack of data-capturing mobility and availability of real-time information hinders the commercialization process and severely limits the applications of this class of gas sensors. To characterize the risks from pollution, there is a great need to develop portable, low-power, and low-cost gas sensors which are able to detect temporal and spatial quantities of multiple pollutants simultaneously at sub ppm-level concentrations over long periods of time. A great advantage of the smartphone-based sensor system we are developing is that it will greatly reduce the overall cost of the system by utilizing smartphones to provide the user interface functionality and computing power for the system. During last few years, UCR has been working on the developing of high density nano gas sensor array using hybrid nanostructures (i.e., single walled carbon nanotubes functionalized with various metals, metal oxides, and conducting polymers) as sensing materials in chemiresistors and chemical field effect transistor (ChemFET). By nanoengineering the sensing materials, we demonstrated a highly sensitivity with fast response and recovery time. In addition, the array is integrated into a housing unit for the final device and linked with a smartphone using Bluetooth wireless technology. The smartphone is running the open source Android platform, allowing for ease of programming to control the device and create a user interface. The smartphone interface is also programmed to collect geospatial data and temperature readings during sensing data collection, to assist in drawing trends from the data and building models. In addition, the ability to load the program onto any Android-based smartphone is free the final control unit from being tethered to a computer, allowing for instantaneous measurements anywhere in the field. During the presentation, the latest development of smartphone integrated high density nano gas sensor array will be presented.

(606c) Metal Oxide Nanowire Sensors and Sensor Arrays for Hazardous Gas Detection

Zhiyong Gu, Chemical Engineering, University of Massachusetts Lowell, Lowell, MA

Nanowires have been widely used for sensor applications due to their unique 1-D structure, high aspect ratio, improved specific surface area and enhanced sensitivity toward certain analytes. In this presentation, I will show our recent effort in developing various metal oxide nanowires as sensors or sensor arrays for hazardous gas detection. A variety of metal oxide nanowires, including tin oxide, copper oxide, doped tin oxides, and mixed tin-copper oxides, have been successfully synthesized by electrodeposition or codeposition method using nanoporous templates. The structure and morphology of these metal oxide nanowires were characterized by SEM, TEM, EDS, and XRD. A novel method has been developed to align the nanowires and integrate them into sensor devices or sensor arrays, using combined dielectrophoretic assembly and thermal oxidation methods. Hazardous gases, including hydrogen sulfide (H_2S) and organic vapors, were tested with these nanowire sensor devices at different concentration levels and elevated temperatures. Copper oxide nanowire sensor has shown promise for H_2S detection with very high sensitivity. For tin oxide based sensors, four types of doped tin oxide nanowire sensors (Pt, In, Ni, and Cu) were consolidated into an array in order to resolve the issue of poor selectivity toward organic vapor detection. Both dynamic and static methods were utilized to obtain the response kinetics and sensitivity data for analysis. Sensing performance was characterized for different organic vapors at a wide range of concentration levels. Finally, principle component analysis (PCA) and pattern recognition technique such as decision tree learning were applied to analyze the classification ability of the sensor array.

(606d) Aptamer Engineering for Ultrasensitive Detection of Pandemic Viruses and Emerging Contaminants

Man Bock Gu, College of Life Sciences and Biotechnology, Korea University, Seoul, South Korea

Young Sup Kwon, Nurul Hanun, Jeewoong Park, Hobin Seo, and Man Bock Gu*

School of Life Sciences and Biotechnology, Korea University, Anam-dong, Seongbuk-gu, Seoul 136-713, Republic of Korea. +equally contributed.

*corresponding author: *mbgu@korea.ac.kr*

Aptamers are single-stranded nucleic acids having molecular recognition properties similar to antibodies, and isolated by in vitro selection and amplification process, SELEX. Aptamer engineering includes innovative screening and modification of aptamers mazimizing their function. This talk will start with how the aptamers are innovatively screened, for the first time in world, by using new nano-material, graphene, without immobilization of targets. A couple of successful examples for protein and pandemic virus using immobilization-free screening of aptamers will be presented (1). Once the aptamers are successfully screened out, the aptamers can be further processed, so called truncation. In this talk, it will be presented how the aptamer candidates are analyzed and heuristically engineered for maximizing its function. This sequence-based engineering of aptamers has resulted in very successful ultrasensitive biosensing, especially for the detection of pandemic virus and antibiotics, which can never been able to be reached without engineering of aptamers. Five different aptamers screened in our lab for small toxic molecules such as oxytetracyclin, tetracyclin, ibuprofen, diclofenac, beta-estradiol were used in this study (2). In addition, the benefits of using nano-sized materials for biosensing will be presented with scientifically clear examples of different hybrid forms of nanomaterials and aptamers, such as aptamers-gold nanoparticle composites. (1) ChemComm (2012), (2) Biosen. Bioel. (2010, 2011).

(606e) Microfluidic in-Channel Growth of 3-D Nanostructures and Their Applications

Joseph Parisi Jr., Chemical, Materials, and Biomolecular Engineering, University of Connecticut, Storrs, CT and Yu Lei, Chemical, Materials & Biomolecular Engineering, University of Connecticut, Storrs, CT

Microfluidic In-channel Growth of 3-D Nanostructures and Their Applications

Joseph Parisi, Yu Lei

Department of Chemical, Materials, & Biomolecular Engineering, University of Connecticut, 191 Auditorium Road, Unit 3222, Storrs, Connecticut 06269

Microfluidic devices have been used in many different applications and have continued to display their ability to shrink large scale experiments into small scale devices. Experiments ranging from cell detectors to microreactors have successfully been implement in microfluidic devices, however there are many areas where the benefits of microfluidic devices have yet to be fully researched. Our work focuses on *in-situ* preparation of 3-D nanostructured electrodes within a microfluidic channel. These nanostructures are fabricated in a fraction of the time of traditional lithography, CVD and hydrothermal techniques, require less material and limits any complicated fabrication steps. The true novelty of the device is in its ability to be immediately ready for in-channel detection of chemical or biological species without the need for alignment steps or mounting the structures on a suitable surface.

(606f) Non-Biological Inhibition Based Sensing (NIBS) for Detection of Trihalomethanes (THMs) in Drinking Water

Isaac K. Afreh, Chemical and Biomolecular Engineering, University of Akron, Akron, OH and Chelsea N. Monty, Department of Chemical and Biomolecular Engineering, The University of Akron, Akron, OH

The use of chlorine for disinfection is a very popular and effective method for drinking water treatment. The chlorine disinfectants are able to destroy waterborne pathogens and also protect drinking water from contamination during delivery to consumers. However, chlorine reacts with natural organic matter (NOM) and anthropogenic chemicals present in the water to form disinfection by-products (DBPs), such as trihalomethanes (THMs), which can be carcinogenic, genotoxic, cytotoxic, and hepatotoxic at moderate concentrations. Unfortunately, there are currently no "real-time" techniques available to monitor for low levels of THMs in finished drinking water. Many drinking water treatment facilities must send samples to off-site laboratories in order to determine DBP concentrations. This can create a delay in detection of up to 48 hours and result in the distribution of potentially toxic drinking water to consumers. Therefore developing a detection method that will rapidly sense low concentrations of regulated THMs in aqueous samples would help mitigate potential health concerns and increase the quality of water being delivered to consumers' homes. Our solution is to develop a sensor that will provide a "real-time" detection of toxic THMs in drinking water. The chemistry for this sensor is based on a new technique in chemical amplification called non-inhibition based sensing (NIBS). In NIBS, instead of the analyte (THM) acting as a catalyst, the analyte rather acts as an inhibitor for the given reaction. The amount of the THM present in the reaction is then determined from the kinetics of the inhibited NIBS reaction. The catalyst solution for the reaction is prepared via Fujiwara reaction, where the THM (inhibitor) binds to the pyridine (catalyst). In this presentation, I will discuss the optimization of the Fujiwara reaction, which will comprise mainly of effects of temperature, residence time and NaOH/ pyridine ratio on Fujiwara reaction. Also, I will expound on the optimization of the catalytic reaction, which will include the effects of reactant concentrations and temperature on the catalytic reaction. Finally, I will demonstrate proof of NIBS concept by showing examples of inhibition versus concentration for the THMs (chloroform, bromoform, dichlorobromomethane and dibromochloromethane).

(606g) A Metabolism-Inspired Assay to Predict Toxicity in Drinking Water Systems

Alvaro A. Rodriguez, Chemical and Biomolecular Engineering, The University of Akron, Akron, OH and Chelsea N. Monty, Department of Chemical and Biomolecular Engineering, The University of Akron, Akron, OH

Water quality analysis is essential in the identification of chemical contaminants in drinking water distribution facilities. In vivo biotransformation of these compounds often generates reactive intermediates that can covalently bind to cells inducing cancer, cytotoxicity and hepatotoxicity. Analytical techniques such as GC/MS and HPLC/MS are frequently used to determine the concentration of toxic compounds offering sensitivity and selectivity. However, such a detailed analysis requires expensive equipment, long processing times and lab personnel with extensive expertise in the analysis of water samples, all of which can limit its use and application. In the case of biological toxicity assays, in vivo or in vitro models involve the manipulation of fragile systems such as bacteria, algae, perfused liver, microsomes, tissue, and cells, yet their use is limited due to shelf life issues. In contrast, bio-inspired chemistries can determine toxicity using more stable compounds. The purpose of this work is to develop a bio-inspired toxicity assay for Phase I and II metabolism by monitoring glutathione (GSH) depletion. Metalloporphyrins are employed as structures that imitate the active site of cytochrome P450 (CYP450) in Phase I metabolism, and Phase II metabolism is carried out by GSH conjugation of metabolites by using a reverse micelle system comprised of cetyltrimethylammonium bromide (CTAB), catalyst that mimics the role of glutathione transferase (GST) in the detoxfication step. Assays of this kind can be utilized to rapidly estimate toxicity in water samples by electrochemically measuring GSH depletion. The fabrication of a microfluidic device will permit the implementation of this technology be used as a portable toxicity assay.

(606h) Sensitive Piezoelectric Cantilever Sensors for DNA, Parasites and Toxins

Kishan Rijal, Sen Xu and **Raj Mutharasan**, Department of Chemical and Biological Engineering, Drexel University, Philadelphia, PA

Abstract. Macro-scale piezoelectric self-sensing and self excited cantilever sensors that exhibit high-order modes near ~ 1 MHz are shown to exhibit sub-femtogram sensitivity in both ssDNA immobilization and hybridization assays in flow format. We show in this presentation that the sensors are able to (1) detect the presence of single nucleotide polymorphism at 1 fM, (2) conduct polymerase chain extension reaction on the sensor, and (3) detect *stx2* gene derived from the pathogen *E. coli* O157:H7 at 400 cells, (4) detect waterborne parasites such as *Cryptosporidium parvum* and *Giardia lamblia*, and toxins such as microcystins. Examination of sensor response in presence of non-target proteins and extraneous material is also characterized.

Introduction. Currently, detection of DNA and RNA oligomers in complex matrix containing extraneous nucleic acid and non-target proteins is of great interest in medical diagnostics. Also of interest is the detection of waterborne parasites and toxins. These applications are as varied as identification of pathogen to tissue dependent microRNA (miRNA) expression. The latter has important medical applications in diagnosis of cancer and degenerative neurological disorders. Resonant-mode cantilever sensors have been shown to be highly sensitive in liquid environments and give real-time measurement of DNA hybridization *in situ* under continuously flowing liquid conditions. We examine application of piezoelectric-excited millimeter-sized cantilever (PEMC) sensors for detecting both natural and synthetic ssDNA and RNA. We also examined detection sensitivity for the waterborne parasites in natural river water and in waters spiked with non-target proteins.

Experimental. Piezoelectric cantilever sensors were fabricated as described earlier[1]. These sensors have been characterized by various means to show sensitivity at femtogram levels[2-6]. A typical detection experiment begins with securing the sensor in a custom isothermal flow cell at a flow rate of ~ 0.5 mL/min and 32 °C. After initial steady-state in resonant frequency is reached the PEMC gold surface was functionalized with thiolated DNA probe and then remaining gold sites were back-filled with 6-mercaptohexanol (MCH) to prevent non-specific adsorption and optimize secondary structure of the functionalized layer for target hybridization. Subsequent to surface functionalization, the circulating buffer was changed to sample containing target nucleic acid (e.g. *stx2* gene derived from *E. coli* O157:H7, others). In some experiments, after the target hybridization step, the four dNTP and *Taq* polymerase was introduced to observe and measure increase in sensor response as the probe ssDNA was extended along the target strand. Experiments with target sequence containing SNP in the middle or at the end of the probe sequence was carried out to determine fidelity in sensor response.

For detecting parasites, source water (tap, river) was spiked with *Cryptosporidium parvum* and *Giardia lamblia* at of 5 to 10,000 (oo)cysts/mL and microcystins LR in the range of 1 pg/mL lo 1 ng/mL in both tap water and then exposed to PEMC sensors functionalized with the corresponding IgG antibodies and installed in a flow cell for in-liquid measurement

Results and Discussion. Typically, ~ 2 mm^2 gold layer is used as the PEMC functionalized surface area. Based on the average diameter of the double helix (~ 2 nm), we estimate that a maximum ~ 10^{12} ssDNA probes (~ 7.4 ng for a 7 KDa ssDNA) can be chemisorbed, assuming a defect free Au-surface. The following are salient results: [1] E. coli O157:H7 at various concentrations were mixed with ground beef mix was used as test natural sample. A short heat-and-chill procedure, and mechanical shearing method was developed to extract genomic DNA, and various concentrations representing from 400 to 5000 cells were prepared, and was exposed to sensors prepared with an immobilized ssDNA strand (designed with Primer criteria) for detecting *stx2* gene. Control experiments with a wild strain of *E. coli* gave no response while the pathogen have over 400 Hz response for the 400 cell sample, and over 1600 Hz shift for sample containing 4,500 cells. [2] A 288-nt ssDNA was first hybridized to the sensor immobilized with an 18-mer ssDNA probe which caused a shift down of 1300 Hz for a target of 3 aM. The position of hybridization left ssDNA containing 103 nucleotides. Subsequent addition of the four dNTPs and Taq polymerase cased a further shift of 1800 Hz indicating the increased in mass due to polymerase reaction. Polymerase reaction was confirmed by lack of hybridization and hybridization with a 20-mer that was complementary to the exposed 103-nt strand. [3]. Hybridization experiments with SNP located in the middle of a 10-mer gave no response, while it being present in the middle of a 20-mer and larger gave a small response of a few hundred Hz. Subsequent exposure to complementary strand gave full hybridization response. [4]. Detection of *C. parvum* was done both in PBS and in 25% milk background at 1 mL/min. The limit of detection was 5 oocysts in 25% milk. Following the detection step, a second IgM antibody specific to *Cryptosporidium* was exposed to the sensor for confirming detection. The response correlated with *C. parvum* concentration tested, and served as a method for reducing false positive sensor response. [5] Detection of *G. lamblia* was carried out at 0.5 mL/min to 5.0 mL/min in a similar manner. Detection of as few as 10 cysts is feasible in 30 min. Higher flow rates gave a higher sensor response magnitude and provided improved signal-to-noise ratio. The specificity of the sensor was challenged with non-target cells (*E. coli* O157:H7 and *C. parvum*). Detection was confirmed with environmental scanning electron microscopy. Successful detection of *G. lamblia* cysts spiked tap water and Schuylkill River (Philadelphia, PA) water showed the feasibility of direct assay of source water samples without a pre-treatment step. One liter samples at 1 cyst/mL or 10 cysts/mL were successfully analyzed.

Conclusion. In summary, we showed PEMC sensors are extremely sensitive and selective, yet robust enough for rapid detection of waterborne parasites. Presence of non-target strands does not interfere with hybridization.
Acknowledgement. The authors acknowledge financial support through EPA STAR Grant R833007 and NSF CBET-0828987.

References

1. D. Maraldo, K. Rijal, G. Campbell and R. Mutharasan, *Anal. Chem.*, 2007, **79**, 2762-2770.

2. G. A. Campbell and R. Mutharasan, *Anal. Chem.*, 2007, **79**, 1145-1152.

3. D. Maraldo, F. U. Garcia and R. Mutharasan, *Anal. Chem.*, 2007, **79**, 7683-7690.

4. D. Maraldo and R. Mutharasan, *Anal. Chem.*, 2007, **79**, 7636-7643.

5. D. Maraldo and R. Mutharasan, *Sens. Actuator B-Chem.*, 2008, **132**, 140-148.

6. K. Rijal and R. Mutharasan, *Langmuir*, 2007, **23**, 6856-6863.

(606i) Plenary Lecture: Biological Sensing Applications of II-VI Semiconductor Nanocrystals

T. J. Mountziaris, Department of Chemical Engineering, University of Massachusetts, Amherst, MA

Nanocrystals of II-VI compound semiconductors (also known as quantum dots or QDs) have unique properties, including size-tunable photoluminescence, narrow emission and broad absorption spectra, high quantum yields, high extinction coefficients, and high photochemical stability, which make them attractive for biological sensing applications that combine detection sensitivity and multiplexing capabilities. This presentation will provide an overview of the state of the art in synthesis, functionalization, and applications of these materials in biological sensors. Research focusing on II-VI nanocrystals that are free from heavy metal ions will be emphasized, including our own group's work on ZnSe-based nanocrystals. Examples of novel QD-based assays for rapid quantitative detection of biomolecular targets, such as DNA, antibodies, and various proteins will be presented. Challenges and opportunities for commercial development of such technologies in the areas of point-of-care diagnostics, emergency care, and environmental monitoring will be discussed.

(618a) Label-Free Analysis of DNA Methylation Using an Optical Resonator

Rasheeda M. Hawk, Mork Family Department of Chemical Engineering and Material Science, University of Southern California, Los Angeles, CA and Andrea M. Armani, Mork Family Department of Chemical Engineering and Materials Science & Ming Hsieh Department of Electrical Engineering-Electrophysics, University of Southern California, Los Angeles, CA

We report a highly specific and sensitive optical biosensor capable of detecting methylated DNA at picomolar levels. 5-methylcytosine (5-mC) and 5-hydroxymethylcytosine (5-hmC) are epigenetic markers involved in gene regulation, cell differentiation, and development. Several techniques have been used to identify and quantify epigenetic markers but many of these techniques are unable to clearly distinguish between 5-methylcytosine and 5-hydroxymethycytosine. Therefore, we immobilized antibodies specific for methylated cytosine and hydroxymethylated cytosine to our resonator microtoroid devices by an epoxy silane which forms a covalent link between the epoxide ring and an amine group on the antibody. This allows for a coupling system that is stable in a buffered saline solution and does not compromise the ability of the resonator. Our functionalized devices are able to differentiate between cytosine, 5-mC, and 5-hmC, offering single nucleotide resolution. The data presented here show that microtoroid resonators can be used in biomedical research and are capable of providing quantitative analysis of DNA methylation in real time.

(618b) Single Molecule Liquid Crystal Detection of Double-Stranded DNA

Stephanie M. Malone[1,2] and Daniel K. Schwartz[2], (1)Chemical Engineering, Carnegie Mellon University, Pittsburgh, PA, (2)Chemical and Biological Engineering, University of Colorado at Boulder, Boulder, CO

The utility of liquid crystals lies in the fact that they orient themselves based on the surface chemistry or topography of a substrate due to elastic constraints. This permits a high level of amplification of surface interactions, which can then be easily characterized with polarized light. Thermotropic liquid crystals were used to explore the extent of surface passivation of glycidoxypropyltrimethoxysilane self-assembled monolayers. Low azimuthal anchoring energy was exploited for the detection of double-stranded DNA. Concurrent polarized and fluorescence microscopy was used to simultaneously observe physical alignment of DNA as well as liquid crystal alignment. The chirality of the molecule, combined with negligible anchoring energy, allowed propagation of the unique alignment of -32±4 when exposed to extended dsDNA. We observed a minimum threshold for sensitivity has been observed to be 200 molecules per μm^2 with theoretical sensitivity as low as 30 molecules per μm^2 . Anchoring energies of double and single-stranded DNA are also compared to further quantify disruption of the director field.

(618c) Multiplexed Nucleic Acid Sensing by Target-Induced Nanoparticle Aggregation with Optical Fiber Cone Arrays

Yunshan Wang[1], Satyajyoti Senapati[1], Li-Jing Cheng[2], Paul Stoddart[3], Scott Howard[4] and Hsueh - Chia Chang[5], (1)Chemical & Biomolecular Engineering, University of Notre Dame, Notre Dame, IN, (2)Chemical and Biomolecular Engineering, University of Notre Dame, Notre Dame, IN, (3)Engineering & Industrial Sciences , Swinburne University of Technology , Hawthorn Victoria, Australia, (4)Electrical Engineering, University of Notre Dame, Notre Dame, IN, (5)Department of Chemical and Biomolecular Engineering, University of Notre Dame, Notre Dame, IN

Environmental monitoring, food safety and disease diagnostics require a rapid and portable sensing platform capable of detecting nucleic acid signatures (DNA/RNA) of multi target pathogens for point-of-care application. Although Real-Time PCR (RT-PCR), a commercial detection kit, can analyse two or more targets in a same reaction, it is expensive, bulky and requires complex data interpretation and hence not suitable for field application. Additionally, detection of targets using RT-PCR requires fluorescence (FL) labeling of primer oligo sequence, which is tedious, time-consuming and limited to target numbers.

To overcome the shortcomings of existing technologies, we report a novel optical fiber sensing platform, which has the potential to detect hundreds of targets in one fiber bundle (1 mm diameter). Each fiber core (~7000 cores in one bundle) can be individually addressed for target analyte sensing using a laser. Several probes (DNA oligoprobes specific to target pathogen of interest) can be functionalized onto one fiber bundle using a photoactivation process. Instead of FL labeling, gold (Au) nanoparticles, that are chemically stable and absorb strongly at certain wavelength, will be used to indicate the presence of target DNA. Target nucleic molecules present in the sample solution act as a bridge between the gold particles functionalized with complementary target sequence and the oligoprobes on the optical fiber array by partially hybridizing with both. In one variation, each optical fiber tip is sharpened by wet etching into a conical shape with a radius curvature less than 100 nm. The conic geometry enhance diffusion transport rate of the nanoparticles as compare to flat surface. Upon capture of gold particles by target DNA on the optical array, an increase in the absorption spectrum confirms the presence of target molecules that can be detected by a portable endoscopy microscope. Further, optical fiber bundle can be easily integrated with a microfluidic channel, where diffusion limitation and sample volume can be optimized. Ion-selective membranes are synthesized near the fiber array to allow analyte concentration by a new depletion zone microfluidic technology (Senapati et al, Topics in Current Chemistry (2011)). Another strategy to reduce the assay time and enhance the sensitivity is to coat the conic tip array with metal and photoconductive material such that plasmonics and optoelectronic dielectrophoresis can be employed to rapidly trap and concentrate the target molecules, as we have done in our earlier biosensor designs (Basuray et al, ACSNano(2009); Cheng et al, LabChip(2010)).

(618d) Real-Time Detection of Micro-RNA Using Cantilever-Based Biosensors

Blake N. Johnson and Raj Mutharasan, Department of Chemical and Biological Engineering, Drexel University, Philadelphia, PA

MicroRNAs (miRNAs) are short non-coding segments of RNA (~ 22 base pairs) recently found to be post-transcriptional regulators of gene expression in higher organisms. Since their discovery, many miRNAs have been identified as important biomarkers of various disease states, such as cancer. Furthermore, they represent a new class of circulating nucleic acids (CNAs) (i.e. those present in flowing bodily fluids). Thus, the ability to measure miRNA in complex liquid matrix at sensitive levels presents an important diagnostic opportunity. We demonstrate the high potential of dynamic-mode piezoelectric cantilever biosensors for making real-time measurements of miRNA (hsa-let-7a) in complex liquid matrix over a large dynamic-range. Binding response is measured by added-mass effect on resonant frequency caused by hybridization between miRNA and immobilized DNA probes. Binding was examined in a variety of liquid formats including pure buffer, serum, and liquid matrix containing a high background of non-target miRNA. Based on concentration-dependent sensor response, limit of detection is competitive with existing technology at sub-picomolar levels.

(618e) Detection of Acetone in Air Using Silver Ion Exchanged ZSM-5 and Zinc Oxide Sensing Films

Kurt L Gerfen, Chemical Engineering, University of Louisville, Louisville, KY, Eric Holt, Chemical Engineering, University of Lousiville, Louisville, KY and **Xiao-an Fu**, Chemical Engineering, University of Lousiville, Lousiville, KY

Zinc oxide (ZnO) based gas sensors with sliver ion exchanged ZSM-5 (Ag^{+}ZSM-5) zeolite overlayer were fabricated and tested for detection of acetone vapor in air. The ZnO and Ag^{+}ZSM-5 films were deposited on 400μm by 400μm platinum interdigitated electrodes (IDE) on a silicon dioxide film on a silicon substrate. Two types of sensors were developed, one with only a ZnO film and the other a ZnO film with Ag^{+}ZSM-5 film on top. The synthesis and materials characterization of ZnO and Ag^{+}ZSM-5 will be discussed. The microfabrication processes used to create the IDE used for the sensors will be presented.

The acetone sensitivity was examined by exposing both sensors to dry air containing varying parts per million acetone at 350°C and then measuring the resulting change in resistance across the electrodes. Both sensors showed sensitivity to acetone in air, however, the sensitivity in the ZnO and Ag^{+}ZSM-5 films were greater than that of just the ZnO film. The utilization of the ZnO and Ag^{+}ZSM-5 films as sensor has great potential for detection of acetone in other gaseous mixtures, including human breath.

(618f) A NOVEL Reactive Chip for Analysis of Carbonyl Compounds in Exhaled Breath

Mingxiao Li[1], Souvik Biswas[2], Michael H. Nantz[2], Richard M. Higashi[2] and Xiao-an Fu[3], (1)Chemical Engineering, University of Louisville, Louisville, KY, (2)Chemistry, University of Louisville, Louisville, KY, (3)Chemical Engineering, University of Lousiville, Lousiville, KY

We report a fabricated reactive chip with thousands of micropillars in a microfluidic channel that have been engineered to selectively preconcentrate trace volatile carbonyl compounds (ketones and aldehydes) in exhaled breath. Certain carbonyl compounds have been identified as biomarkers of lung diseases including lung cancer, Tuberculosis, and cystic fibrosis. The chips are fabricated using deep reactive ion etching (DRIE) of silicon. The micropillar surface is functionalized with a reactive quaternary ammonium aminooxy salt or a reactive neutral aminooxy compound for trapping trace carbonyl compounds by means of an oximation reaction. The chip can be used for directly preconcentrating carbonyl compounds in breath. The concentrated ketone- and aldehyde-ATM adducts are analyzed by Fourier transform-ion cyclotron resonance (FT-ICR) mass spectrometry (MS) and gas chromatography-mass spectrometry (GC-MS). The concentrations of all ketones and aldehydes in breath can be detemined by FT-ICR-MS and GC-MS. We demonstrate the utility of this novel chip in the detection of C1- to C12- ketones and aldehydes in exhaled breath.

(618g) Novel Biosensors for Bioprocess Development

Alireza Behjousiar, Chemical Engineering and Chemical Technology, Imperial College London, London, United Kingdom

<u>**Novel biosensors for bioprocess development**</u>

It has long been a goal of the bioprocessing field to be able to produce proteins in a cost effective and efficient manner. The most basic stage of protein production is to optimize the growth of the cells in question to give higher and more specific protein production. The current method of choosing high producing stable mammalian cells is labour intensive and time consuming, representing an opportunity to employ new analytical methodologies in an effort to expedite progress. The established cell line construction and selection methods rely heavily on achieving high cell growth rates, assessed in terms of confluence, in the early stages of cell line development. We suggest the use of genetically encoded FRET biosensors, which will allow us to use a low volume assay at an earlier stage in cell selection to measure key metabolites in cell culture as signs of well being of these cells.

The proposed system is able to monitor the concentrations of key intracellular metabolites including glucose and glutamine in real time, in a non-invasive way. Glucose and glutamine are key energy sources for cultured cells, and therefore vital metabolites to monitor in cell line selection. Therefore for efficient cell growth to occur glucose and glutamine must be present in a steady supply and not used up to quickly. The system consists of CHO-S cells stably transfected with biosensors for glucose/glutamine, which measure metabolites in the relevant range for healthily growing cells. The fluorescent signal of these biosensors has been calibrated against intracellular glucose and glutamine concentration measurements to produce *in vivo* calibration curves (Behjousiar *et al., 2012*).

We have further established the use of these biosensors for bioprocess development in a high throughput culture system (BioLector, m2p labs). Using cells transfected with the glutamine biosensor we performed a medium optimisation experiment using CD-CHO and different amounts of L-glutamine *versus* two different types of glutamine containing dipeptides. We monitored pH, DOT, and glutamine utilisation in 48 well plates and assessed the growth kinetics, integral viable cell concentration, and glutamine utilisation rates.

The proposed system for real-time, *in situ* monitoring of intracellular metabolites will allow us to make informed decisions regarding which cell lines to progress to subsequent stages of selection. Importantly, it will enable the selection of cell lines that present better characteristics in terms of manufacturability at an earlier stage than conventional methods. Our work also presents data to support the usefulness of the BioLector as a valid tool for cell maintenance and metabolic monitoring in an automatic, high throughput manner.

(618h) Monitoring of Conformational Change of a Calcium Binding Protein S100A5 Using QCM-D Sensor

Hyun J. Kwon, Engineering and computer science, Andrews University, Berrien Springs, MI

Detecting conformational change of protein-protein binding is important in understanding their physiological functions. Calcium dependent proteins undergo various structural rearrangements during binding with various binding moieties. Such conformational change has been measured by NMR, AFM, or calorimetric method. We demonstrate QCM-D characterization of conformational shift in calcium-dependent S100A5 protein for binding to calcium, calmodulin (CaM), olfactory marker protein (OMP), and in its dimerization. Simultaneous measurement of frequency and dissipation change of the QCM-D sensor enables one to characterize film thickness and flexibility upon bindings. His-tagged recombinant S100A5 was immobilized on to the gold QCM surface via NTA-Ni+ and binding moieties were applied. Binding of calcium caused releasing of entrapped water making the thickness of S100A5 layer decrease. Dimerization with non-his tagged recombinant S100A5 caused increased thickness at affinity of 3.7 ± 2 uM. Binding to CaM and OMP caused decreased adlayer thickness with unique characteristics as demonstrated by changes in dissipation vs. frequency plot. The result demonstrated QCM-D sensor could be employed as an effective and simple way to detect diverse binding characteristics of protein conformational changes.

(680a) Developing a Magnetic Microfluidics Setup to Capture Bacterial Contaminants in Blood Products

Jillian Larsen[1], Srigokul Upadhyayula[2], Kenny Chau[3], Vicente Nuñez[2] and Valentine I. Vullev[4], (1)Bioengineering, UC Riverside, Riverside, CA, (2)Bioengineering, University of California, Riverside, CA, (3)University of California, Los Angeles, CA, (4)Department of Bioengineering, University of California, Riverside, Riverside, CA

Since bacterial contamination still remains a risk in transfused blood products, there is a critical need for early and sensitive detection methods. The aim of this project is to develop a high-throughput bacteria-sensing device by incorporating magnetic trapping in microfluidic devices. The immediately available systems for real-time detection are deficient in their Boolean natures, excessive time and reagent consumption, and limited sensitivity in the overall scope of detection. Surface-modified paramagnetic beads facilitate adhesion to bacterial agents flowing through the microfluidic channel. Since a magnet accompanies the microchip, our proposed design of the chip will have the ability to immobilize cells adhered to the paramagnetic beads for analysis. By removing the magnet we are able to wash the trapped pathogens thereby preparing the device for the next detection cycle. This 'recycling' feature of our proposed chip design allows for continuous sampling of bacterial cells for immediate detection. Moreover, magnetic trapping in microfluidic systems will drastically improve the sensitivity of our technique by effectively eliminating the minimum-bacteria-concentration requirement. Coupling magnetic trapping and surface modification of beads in microfluidic devices allows us to image and differentiate the staining kinetics on a massively parallel single-cell level, thereby enabling the ability to continuously sample heterogeneous mixtures of bacterial species. This research will help us develop tools with the ability to couple the low sensitivity of current detection method with the critical time requirement for accurate identification of various bacteria.

(680b) Isolation and Capture of Circulating Tumor Cells by a Microfluidic Device with Functionalized Planar Structures

Hyeun Joong Yoon[1], Tae Hyun Kim[2], Zhuo Zhang[1], Trinh M. Pham[1] and Sunitha Nagrath[1], (1)Department of Chemical Engineering, University of Michigan, Ann Arbor, MI, (2)Department of Electrical Engineering and Computer Science, University of Michigan, Ann Arbor, MI

Circulating tumor cells (CTCs) have been identified in the art in peripheral blood from cancer patients and are likely the origin of metastatic disease. Isolation and capture of CTCs represent a potential alternative to invasive biopsies during diagnosis of disease. In the recent past, various microfluidic based CTC isolation technologies have been developed to capture, indentify, sort, and enumerate CTCs. Micro-sized structures functionalized with anti-epithelial-adhesion-molecule antibodies (Anti-EpCAM) is the common approach to capture CTCs. Here we present a microfluidic device with functionalized planar structure to capture CTCs. The microfluidic device has a PDMS chamber and patterned gold posts. It is much easier to scan, count and grow captured cells in comparison with previous micro-sized post-based structure. CTCs were isolated by assembling Graphene oxide on the planar structures for enhanced sensitivity. The functionalized graphene oxide has been used to selectively capture MCF-7 breast cancer cells with a capture rate higher than 90 %.

(680c) PCR-Independent, Reagent-Free, Binary-Mode Nucleic Acid Detection

Leyla Esfandiari, Bioengineering, Unversity of California Los Angeles, Los Angeles, CA, Jacob Schmidt, Bioengineering, University of California Los Angeles, Los Angeles, CA and **Harold G. Monbouquette**, Chemical and Biomolecular Engineering, University of California, Los Angeles, Los Angeles, CA

A new concept for specific nucleic acid (NA) sequence detection that does not rely on polymerase chain reaction (PCR) for target sequence amplification and does not require any special reagents other than a complementary sequence capture probe conjugated to polymer spheres has been demonstrated. Recent work has shown that the presence of single macromolecules in a nanopore causes measurable changes in the pore's electrical resistance. Our concept takes this single-molecule detection system a significant step further by amplifying the signals from specific single molecule targets to generate an easily detected on/off current signal due to a very large sustained increase in pore resistance. The key elements of the proposed system include a peptide nucleic acid (PNA) capture probe conjugated to polymer spheres. PNA oligomers are uncharged analogs to DNA and RNA that share the same base chemistry and hybridize strongly to complementary NA sequences. Since the sphere-PNA conjugates carry little or no charge, they do not exhibit electrophoretic movement in response to a steady, DC field imposed through a pore. However, the substantial negative charge acquired upon capture of a target DNA or RNA sequence makes the hybridized conjugate electrophoretically mobile. If the pore size tapers to a diameter smaller than the spheres, the charged conjugate carrying the hybridized PNA and target NA would be expected to enter the large end of the pore, block it, and significantly increase its resistance, thereby causing a very strong, sustained drop in measured current. In such a way, this system is expected to give an essentially binary response signaling the absence or presence of a target NA. To date, the concept has been demonstrated at the microscale. In so doing, we have successfully detected ssDNA with a 12-mer portion of a gene encoding the anthrax lethal factor. This proposed new technology would be useful for applications where determination of the presence or absence of NA of a particular sequence, rather than its concentration, is of primary concern such as in patient screening during epidemics, oncological status assessment during surgery, detection of food contaminants, and biowarfare agent detection.

(680d) Integrated Palladium Reference Microelectrode for Use in Electrochemical Detection of Bacterial Toxins

Thaddaeus A. Webster and **Edgar D. Goluch**, Department of Chemical Engineering, Northeastern University, Boston, MA

The miniaturization and integration of components for electrochemical detection in micro- and nano-fluidic devices is of great interest as it directly yields an electrical signal that provides information about the chemical composition of a sample. We have developed a device that integrates a palladium and a gold electrode inside of a nanoscale constriction for use as an electrochemical sensor. We first tested the stability of the palladium pseudo-reference electrode and compared its performance against a commercially available external silver/silver chloride reference electrode, using the gold as a working electrode.

Palladium is capable of absorbing hydrogen up to several hundred times its volume. It functions in a similar fashion to the standard hydrogen reference electrode in that the potential is set by the hydrogen ion concentration versus the amount of absorbed hydrogen molecules. The potential remains constant as the amount of absorbed hydrogen changes over a wide range of values. The potential is affected by the concentration of protons, but for a buffered solution, this also does not change significantly, making it very useful in biological experiments.

We used the device to measure pyocyanin production by *Pseudomonas aeruginosa* in growth media. The oxidation potential of pyocyanin was measured to be approximately -0.5 volts against the miniaturized palladium reference. By utilizing square wave voltammetry, the concentration of pyocyanin was selectively measured in a complex solution.

(680e) Novel Nickel Porous Nanoparticles and Their Application for the Electrochemical Detection of Serotonin

Molly Clay[1], Stephen DeWitt[2], Qingzhou Cui[1], Julie Chen[3] and Zhiyong Gu[1], (1)Chemical Engineering, University of Massachusetts Lowell, Lowell, MA, (2)Chemical Engineering, University of Massachusetts Lowell, Lowell, (3)Mechanical Engineering, University of Massachusetts Lowell, Lowell, MA

Sensing is vital to the healthcare industry in order to diagnose patients with accuracy and in a timely manner. Among many sensing platforms, electrochemical sensing is a fast, reliable, sensitive, and low cost method to improve conventional biosensing techniques. Herein we show the electrochemical detection of serotonin, a neurotransmitter linked to psychiatric diseases including depression and anxiety, obsessive-compulsive disorder, and addiction, using novel porous nickel nanoparticles. These porous nanoparticles are synthesized by a galvanic replacement reaction between an aluminum nanoparticle template and nickel metal ions. The sensor set-up consists of a 3-electrode cell with working, reference (Ag/AgCl), and counter (Pt) electrodes. A glassy carbon electrode modified with porous nickel nanoparticles functions as the working electrode. Due to the increase in surface area, these porous nanostructures have promise as electrocatalysts. It was found that these nickel porous nanoparticles enhance the sensitivity of the electrode towards the oxidation of serotonin with a detection limit of about 25 nM.

(680f) Development and Modeling of Single Particle SERS Assays

Ashley J Driscoll and Patrick A. Johnson, Chemical & Petroleum Engineering, University of Wyoming, Laramie, WY

Nanoparticles have been synthesized from a variety of organic and inorganic materials and find use in fields as diverse as electronics, drug delivery, and bioassays. Nanoparticle synthesis and surface modification provide a simple and reproducible method of preparing nanoscale surfaces. Bioassays for the detection of DNA, antibodies and antigens have benefitted from the application of nanoparticles and the introduction of nanoparticles has also accelerated the use of surface enhanced Raman spectroscopy (SERS) as a bioassay detection method. Gold surfaces provide for facile attachment of biomolecules and gold plasmonic effects in the visible wavelength region make nanoparticles ideal for surface enhanced Raman spectroscopy (SERS) bioassays. Single particle bioassays can be developed by incorporating magnetic capture and a gold functionalized surface into a single nanoparticle. By using methods such as the polyol process, a magnetic cobalt core can be reduced in a diol solution and a gold shell can be deposited subsequently. This produces a particle with the magnetic characteristics of cobalt, and the light scattering and biocompatible character of gold. Single particle assay systems can simplify the collection and reading of target analytes in bioassays by allowing the collection of surface bound analyte from solution by the application of a magnetic field and the formation of a pellet with the particles for detection. When coupled with Raman spectroscopy, only the analyte bound with a gold particle will produce a detectable signal due to the SERS effect.

Gold colloids can provide a large surface area per unit volume and a mobile capture surface that reduces average diffusion distances for analytes in a bioassay solution. Brownian motion of both the analyte and capture substrate also renders the assay less mass transport limited than surface based assays. Surface assays have benefitted from models which elucidate the important physical parameters and their interrelationships, leading to reduced assay time, improved signal generation and improved limits of detection. To further the development and understanding colloidal assays, modeling of the colloidal system could also produce similar gains by improving understanding of the system as well as improving performance, particularly single particle bioassays using surface enhanced Raman spectroscopy (SERS) for detection. We present the preliminary modeling of solution based colloidal assays.

(680g) Polymerization Amplification for Thermal Biodetection

Leila Safazadeh[1], Vivek Balasubramaniam[2], Richard E. Eitel[1] and **Brad J. Berron**[1], (1)Chemical and Materials Engineering, University of Kentucky, Lexington, KY, (2)Univeristy of Colorado at Denver School of Medicine, Aurora, CO

Thermochemical sensors operate by monitoring the thermal energy change resulting from a chemical reaction and represent a simple, inexpensive and reliable method of monitoring a chemical reaction. Biosensors can be designed around thermochemical sensors through the specific biological binding of a critical reaction component to an immobilized phase, introducing the remaining reaction species to the immobilized reagent, and monitoring the temperature, which can be correlated to the concentration of the immobilized component. Most commonly, glucose oxidase is bound to an immobilized phase specifically in the presence of the target biomaterial. Subsequent exposure to an appropriate substrate solution drives a chemical process where a specific quantity of heat is liberated for each reaction in the enzymatic cycle. A sufficiently large change in thermal energy will alter the local temperature, and the temperature may be measured by a variety of reliable, proven methods. For this reason, thermal biosensors have been often promised as a simple, reliable analytical tool. Critically, the traditional, enzymatically-driven thermal biosensor is fundamentally limited by the net heat generated by glucose oxidase.

We describe the use of enzyme mediated polymerization to facilitate the sensitive, quantitative response to a biorecognition event. In this approach, glucose-oxidase is coupled to the target species though specific protein binding, and the enzyme participates in a reaction cascade yielding hydroxyl radicals. Enzyme kinetics dictate the timescale of polymerization, and ultimately enable the quantitation of the target species. The simplicity of the response enables simultaneous quantitation of multiple targets using inexpensive, off-the-shelf technology.

(680h) Dual Delay Line SAW Sensor Utilizing Multistrip Coupler for Liquid Biosensing Applications

Mandek B. Richardson, Chemical & Biomedical Engineering, University of South Florida, Tampa, FL, Changbao Wen, Electronics and Controls Engineering, Chang'an University, China and Venkat R. Bhethanabotla, Chemical Engineering, University of South Florida, Tampa, FL

Dual delay line SAW sensor utilizing multistrip coupler for liquid biosensing applications

Mandek Richardson[1], Changbao Wen[2,3], and Venkat R. Bhethanabotla[1]

[1]*Sensors Research Laboratory, Department of Chemical & Biomedical Engineering, University of South Florida, Tampa, Florida 33620, USA*

[2]*Institute of Micro-nanoelectronics, School of Electronics and Control Engineering, Chang'an University, Xi'an 710064, People's Republic of China*

[3]*Institute of Vacuum Microelectronics, School of Electronics and Information Engineering, Xi'an 710064, People's Republic of China*

Surface Acoustic Wave (SAW) devices can be used as direct, label free biosensors that monitor the interaction between a receptor and its target in real time, through changes in the properties of the traveling wave (i.e., frequency, phase, and amplitude) [1]. A typical SAW device consists of two metal interdigital transducers (IDTs) separated from each other where one transducer is used to generate acoustic waves and the other receives acoustic waves and converts them into electrical signals. SAW IDTs not only generate surface waves but also bulk waves that have deleterious effects on sensor performance. For example, the generation of bulk waves reduces SAW power, which increases insertion loss, and interference between SAW and bulk waves causes amplitude and phase distortions that increase ripples in the pass band, which affect linear sensor calibration [2, 3]. A technique used to address problems stemming from bulk wave generation was to implement a multistrip coupler into SAW filter designs [2]. An MSC consists of an array of identical electrodes oriented parallel to the surface-wave wavefront; in which, the electrodes are not electrically connected. The purpose of the MSC is to divert the surface wave to an output IDT that is offset from the input IDT; this decouples SAWs and bulk waves allowing just the SAW to arrive at the output IDT. For practicality, SAW substrates with large electromechanical coefficients (K^2) must be used [2].

In this work we present a dual delay-line SAW liquid sensor with weighted input/output IDTs and two MSCs fabricated on a 41° Y-X $LiNbO_3$ piezoelectric substrate. We choose 41° Y-X $LiNbO_3$ because of its high acoustic velocity (4792 m/s), large electromechanical coupling coefficient, K^2 (17.3%), and leaky wave propagation [4]. Substrates that support leaky wave modes can be used as liquid sensors because particle displacement is parallel to the surface, which limits the dissipation of energy into the adjacent liquid. Incorporation of the dual-delay line structure coupled with MSCs eliminates spurious responses caused by the external environment and bulk wave generation [5]. Furthermore, the presence of MSCs improve sensor calibration by reducing the amount of phase distortion in the pass band through the suppression of bulk waves, and they increase sensor sensitivity by trapping the leaky wave at the surface. In addition, weighted IDTs were used as the input and output transducers to improve side lobe rejection levels [5]. Devices based on the design were fabricated and compared to a standard SAW dual delay-line to show its feasibility as a biosensor.

References

[1] M. Bisoffi, B. Hjelle, D. C. Brown, D. W. Branch, T. L. Edwards, S. M. Brozik, V. S. Bondu-Hawkins, and R. S. Larson, "Detection of viral bioagents using a shear horizontal surface acoustic wave biosensor," *biosensors &*

Bioelectronics, vol. 23, pp. 1397-1403, 2008.

[2] C. Campbell, *Surface acoustic wave devices for mobile and wireless communications / C.K. Campbell*: San Diego : Academic Press, c1998., 1998.

[3] D. W. Branch and T. L. Edwards, "Love Wave Acoustic Array Biosensor Platform for Autonomous Detection " *IEEE Ultrasonics Symposium Proceedings,* pp. 260-263, 2007.

[4] K. Lee, W. Wang, T. Kim, and S. Yang, "A novel 440 MHz wireless SAW microsensor integrated with pressure-temperature sensors and ID tag," *Journal of Micromechanics and Microengineering,* vol. 17, pp. 515-523, Mar 2007.

[5] C. Wen, C. Zhu, Y. Ju, H. Xu, and Y. Qiu, "A Novel Dual Track SAW Gas Sensor Using Three-IDT and Two-MSC," *IEEE Sensors Journal* vol. 9, pp. 2010-2015, 2009.

(736a) CeO2 Nanofibers for in-Situ O2 and CO Sensing in Harsh Environment

Yixin Liu [1], Lichun Zhang[2], Puxian Gao[3] and Yu Lei[1], (1)Chemical, Materials & Biomolecular Engineering, University of Connecticut, Storrs, CT, (2) Institute of Materials Science, University of Connecticut, (3)Institute of Materials Science, University of Connecticut, Storrs, CT

As the most common type of harsh environment sensors, high temperature gas sensors are paramountly important to improve combustion efficiency and control emissions. In this study, large scale and polycrystalline CeO_2 nanofibers were prepared using a facile two-step synthesis route including electrospinning and calcination. The structure and composition of nanofibers were characterized by SEM and X-ray diffraction.The as-prepared CeO_2 nanofibers show good morphological and structural stability at high temperature environment (800~1000 ^oC). CeO_2 nanofibers were further employed for *in-situ*, real-time oxygen (O_2) and carbon monoxide (CO) sensing at 800 ^oC and 1000 ^oC, respectively, both of which showed sensitive, reversible, and reproducible response. To the best of our knowledge, the good sensitivity of the sensor is one of the best results for resistor-type sensor for CO sensing at such high temperature.The mechanism for distinguishable response towards O_2 and CO was discussed. These results indicate that electrospun CeO_2 nanofibers are a promising nanomaterial for gas sensing in harsh environment.

High Temperature Potentiometric O_2 Sensor Based On Stabilized Zirconia and NiO Electrode

Xiangcheng Sun[1], Haiyong Gao[2], Yixin Liu[1], Puxian Gao[2] and Yu Lei[1],

(1)Chemical, Materials & Biomolecular Engineering, University of Connecticut, Storrs, CT, (2)Institute of Materials Science, University of Connecticut, Storrs, CT

Abstract

A planar O_2 potentiometric sensor was fabricated by using yttria-stabilized zirconia (YSZ) electrolyte, NiO sensing electrode and Pt reference electrode. The thin NiO films (or NiO nanofibers) of 100 nm thickness were prepared by r.f. magnetron sputtering of NiO on YSZ substrate (or by electrospinning on YSZ substrate followed by calcination). X-ray diffraction patterns suggested that NiO was coated on YSZ surface. The morphology of NiO film (or NiO nanofiber) was investigated by scanning electron microscopy and atomic force microscopy. The sensing electrode was examined for O_2 sensor operating at 500-700°C. The linear relationship between the sensor EMF output and the logarithm of the O_2 gas concentration was observed, and the sensitivity to oxygen increased with increasing operating temperature. These results obeyed the theoretical Nernst relationship well. Finally, electrochemical impedance spectroscopy study was used to investigate the sensor mechanism, and a novel potentiometric sensor mechanism which involves surface reaction, catalysis, oxidation and reduction was suggested.

Introduction

High temperature sensors used in combustion processes can help for real time monitoring, and fundamental studies of combustion processes. Thus, this kind of sensors would be helpful for toxic emissions reduction and energy efficiency improvement in combustion processes. The development of high temperature gas sensors is highly desirable and has the potential to serve a large market. Due to the importance of oxygen in combustion environment, this research will focus on high temperature oxygen gas sensors.

To detect oxygen, a multistage configuration device with external oxygen reference is the most common approach. In this design, air is provided from outside the combustion environment to a reference electrode that is separated from the sensing environment. The oxygen concentration difference between the reference air and the measuring environment generates an open circuit potential. Therefore, the concentration of oxygen can be calculated. However, oxygen leaking from the reference chamber is always a problem in high temperature oxygen sensor for commercial application. In addition, sensor configuration was complex, and sensor cost was relatively high. Therefore, a new kind of simple, and low cost oxygen sensor is highly desirable.

So far, the mixed-potential type sensors have attracted much attention due to their excellent performance in high temperatures. Mixed potential type sensors develop open-circuit electromotive force due to the difference in kinetics of redox reactions of various gaseous species at each electrode/electrolyte/gas interface. Compared with other electrochemical sensors, mixed potential type sensors have advantages such as simple sensor configuration, mechanical thermal and chemical stability and the potential of working at high temperatures.

Electrochemical impedance spectroscopy (EIS) is a powerful technique that have been used to understand and distinguish electrochemical processes that occur on solid state electrochemical fuel cells and sensors. EIS can be applied to study the effects of independent variables like temperature, microstructure, and gas concentration on different system processes. EIS has the extraordinary ability to access system time constants ranging from 10^{-7} to 10^{3} s. Through time constants study by EIS that contributes to different parts, response time issues of sensors can be refined. In addition, theoretical analysis from impedance data would be helpful for discerning sensor mechanisms.

In this research, a novel planar configuration sensor using NiO as sensing electrode was fabricated and then oxygen sensing performance at different temperatures was investigated, sensing mechanism through electrochemical impedance study was suggested.

Materials and Methods

The side view of planar sensor configuration was shown in Fig. 1. The sensor was fabricated through using a YSZ plate (8 wt % Y_2O_3-doped, 6×8 mm, 0.5 mm thickness, Coating & Crystal Technology Inc.). On one side of YSZ substrate, the thin NiO film of 100 nm thickness was prepared by r.f. magnetron sputtering. Pt paste printed on the other side of YSZ substrate worked as reference electrode. After that Pt wires were pressed on both sides of YSZ electrolyte, and connected to a CHI 660D electrochemical analyzer (CH Instruments Inc., USA) through two long Ni–Cr alloy wires. The sensor was placed in a furnace with a temperature control to conduct high temperature gas sensing. Impedance spectra were performed through CHI in the frequency of 0.1 Hz to 1 MHz with 100 mV rms perturbation.

Fig. 1 Side view of planar sensor configuration

Results and Discussion

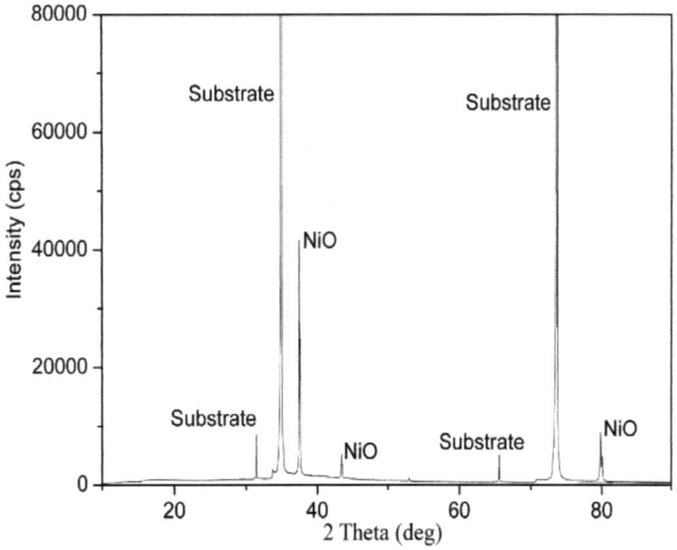

Fig. 2 XRD patterns of NiO films prepared by sputtering after 1200°C sintering

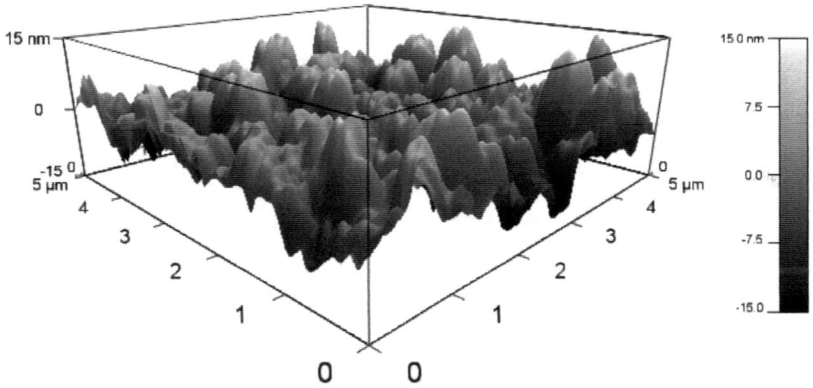

Fig. 3 Top view of AFM graph of NiO film after 1200°C sintering

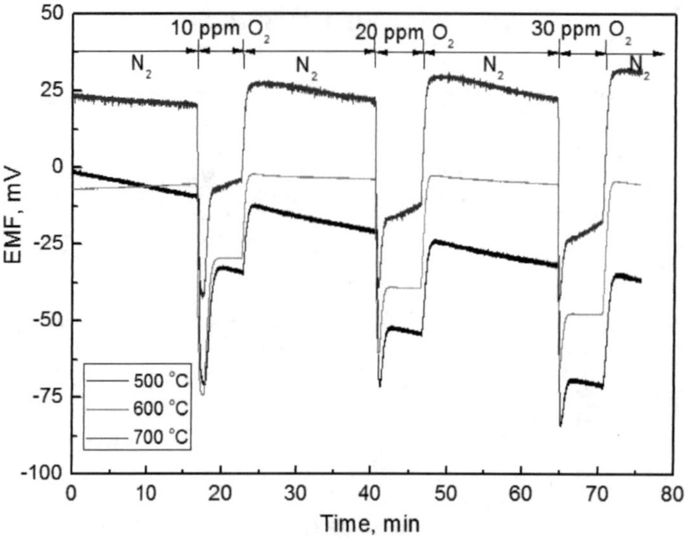

Fig.4 In-situ sensor response to various oxygen concentrations at different operating temperatures

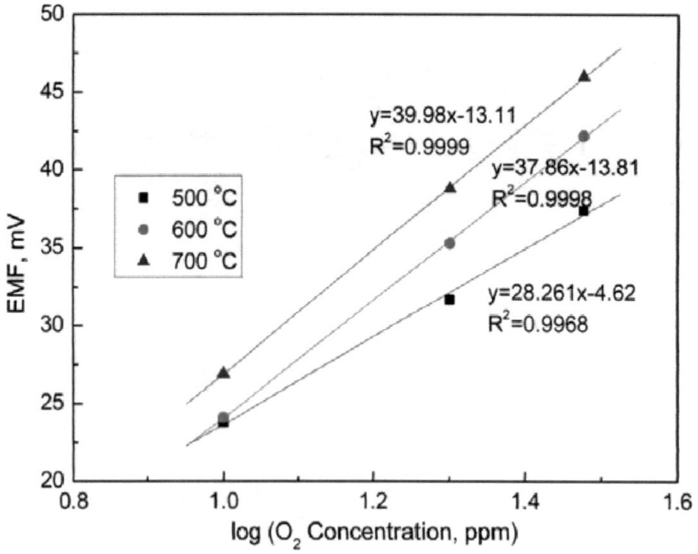

Fig. 5 Oxygen sensitivity of planar potentiometric sensor at different temperatures

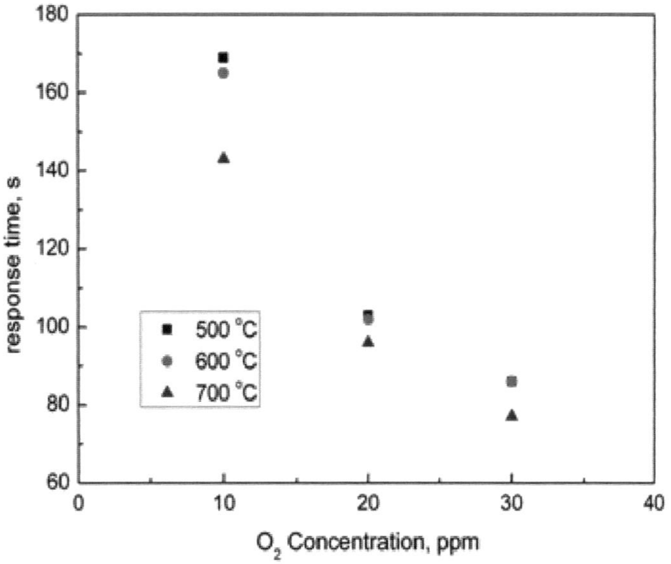

Fig. 6 Response time of potentiometric sensor to various O_2 concentration at different temperatures

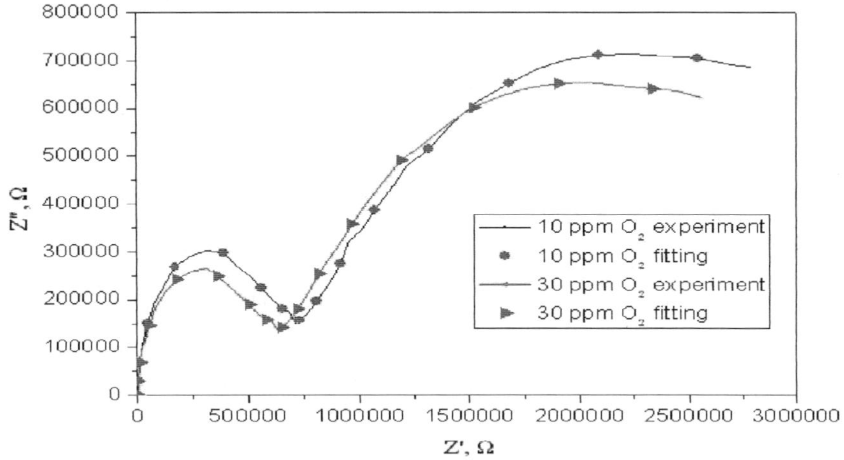

Fig. 7 Complex impedance plots in 10 ppm and 30 ppm O_2 at 600 °C for the sensor

Conclusions

A novel planar potentiometric sensor using NiO as sensing electrode had been proposed, which had simple configuration. The EMF output of sensor varied logarithmically with O_2 concentrations, and sensitivity to O_2 increased with increasing of temperature, from 28.26mV/dec at 500°C to 39.98mV/dec at 700°C. These results fitted well with Nernst Law. In addition, with increase of gas concentration, senor response time greatly decreased, while temperature influenced a little on response time. This suggested mass transport might control sensing process. EIS was used to study this process, two semicircles were observed. Low frequency semicircle which related to oxygen diffusion and transfer was much larger. The result suggested oxygen diffusion and transfer had a greater impact on sensing process. The refined sensing mechanism of the present potentiometric sensor needs further investigation.

OPTICAL SENSOR FOR NANOPARTICLE IDENTIFICATION IN LIQUIDS

Jarkko J. Saarinen[1,2,], Jun Uozumi[2], Erik M. Vartiainen[3], and Kai-Erik Peiponen[4]*

[1] *Laboratory of Paper Coating and Converting, Center for Functional Materials, Abo Akademi University, Turku, FINLAND*
[2] *Faculty of Engineering, Hokkai-Gakuen University, Sapporo, JAPAN*
[3] *Department of Mathematics and Physics, Lappeenranta University of Technology, Lappeenranta, FINLAND*
[4] *Department of Physics and Mathematics, University of Eastern Finland, Joensuu, FINLAND*
[*] *Corresponding author: jarkko.j.saarinen@abo.fi*

Abstract

A method for optical identification of dielectric and metal nanoparticles in a liquid matrix is presented using the maximum entropy method (MEM) for phase retrieval of reflectance with tranverse electric (TE) and tranverse magnetic (TM) linearly polarized light. The effective dielectric function of the colloid is obtained using phase retrieved complex reflectance components. An excellent accuracy is observed both for dielectric and metallic nanoparticles with volume fractions up to 10 percent. It is believed that such nanoparticle identification will find applications in monitoring systems for identification of nanoparticles and their concentration in water streams.

Introduction

The study of optical properties of colloids has a rather long history dating back to the start of the 20th century and works by Maxwell Garnett who studied colors of glasses containing minute metal spheres [1]. He also presented an elegant theory to predict the optical properties of such effective media that is even nowadays widely used in the description of optical properties of colloids. The spectroscopic properties of colloids have been important in basic research of colloidal physics and chemistry since the optical properties of a colloid depend on the size, shape and structure of the nanoparticles, their complex refractive index, and thermodynamic condition of the ambient host liquid. Much of the research work has been on fundamental optical properties of nanoparticles in liquid environments.

The emergence of nanotechnology and addition of nanoparticles into existing consumer products to improve their performance has extended applications of colloid studies to completely new fields such as nanomedicine and nanotoxicology. For example, colloidal gold nanoparticles have been studied for cancer diagnostics and therapy, and their optical properties such as surface plasmon resonance (SPR) and Raman scattering are important in molecular specific diagnostics and detection of the cancer [2]. Another recent example is addition of functional materials including nanoparticles in food, and issues of nanotoxicology in liquid have been raised [3]. The fundamental properties of such colloids are the intrinsic optical properties of both the host medium and the nanoparticles: the optical properties of the host medium are typically well-known and free of spectroscopic resonances in the frequency range of interest, for example, as in the case of water. On the other hand, nanoparticles can usually be identified with their optical fingerprints. Furthermore, these fundamental properties are independent of the optical process used to access them.

Traditionally surface plasmon resonance has been utilized to study optical properties of liquids and colloids. One can construct an active system *i.e.* a sensor for monitoring of minute change of refractive index of gases or liquids [4, 5]. In the past such sensors have typically utilized a prism with one face coated with a thin metal film such as gold or silver. The prism face with metal coated film is in contact with the liquid to be measured. A traditional method is to use a laser or a LED as a light source, and a polarizer to produce TM-polarized probe light that is needed to excite surface plasmons, which are collective oscillations of the free electrons. Either by changing the angle of incidence or by focusing the incident light beam it is possible to obtain a dip in the reflectance due to the SPR. As a detector one can use a photodiode, array of photodiodes or a CCD-camera. Different kinds of SPR sensors have been constructed and are commercially available. In addition to angular interrogation it is possible to record a reflection spectrum with the aid of a SPR sensor [6]. In such a case one has to scan the wavelength of the incident light. This provides an interesting option because in addition to the SPR dip one is able to get spectral fingerprints of the liquid. Unfortunately, the thin metal film used to excite the SPR may oxidate in ambient conditions and also contaminate.

In this study we show that it is possible to obtain the optical properties of both insulating and conducting nanoparticles from the angle-, wavelength-, and polarization-dependent reflectance, which is *directly* measured from the colloid. This allows us to identify particular nanoparticles and/or to get their concentration in a colloid. In practical measurements one can utilize a polarization maintaining single mode fiber to measure the reflectance at various angles from a colloid containing nanoparticles. A great benefit of reflectance measurement compared to traditional transmission measurement is significantly reduced scattering providing a more accurate identification of nanoparticles.

Here we study the reflectance from the air/nanoparticle colloid interface with spherical dielectric or metal nanoparticles dispersed in a water matrix. From reflectance measurement we get the amplitude of the reflectance but the phase information is lost. Traditionally such phase retrieval problems in reflection spectroscopy have been analyzed using the Kramers-Kronig type dispersion relations [7] that relate the phase and the amplitude of reflectance to each other. Unfortunately, there are some serious drawbacks in the classical dispersion theory analysis as the integral relation nature of the analysis is based on semi-infinite integration from zero to infinity. This, in principle, requires knowledge of reflectance data over the semi-infinite spectral range. However, in reality, we are always limited to finite spectral range. Therefore, we will utilize here the maximum entropy method (MEM) [8]. In the MEM the entropy is a measure of the uncertainty in the data, and hence, the maximum entropy estimate is the least biased estimate on the given data. The method has been successfully applied in various fields ranging from enhancement of astronomical images [8] to predict atmospheric CO2 content [9].

Methods

A Maxwell Garnett (MG) type effective medium theory [1] is utilized here to describe the effective dielectric function of colloids with spherical nanoparticles. We assume that the dielectric nanoparticles have a dielectric function given by a single resonance Lorentzian lineshape function [7] whereas bulk properties of gold [10] are used for the gold nanoparticles. The dielectric function of non-absorbing water host is given by Partington's formula [11].

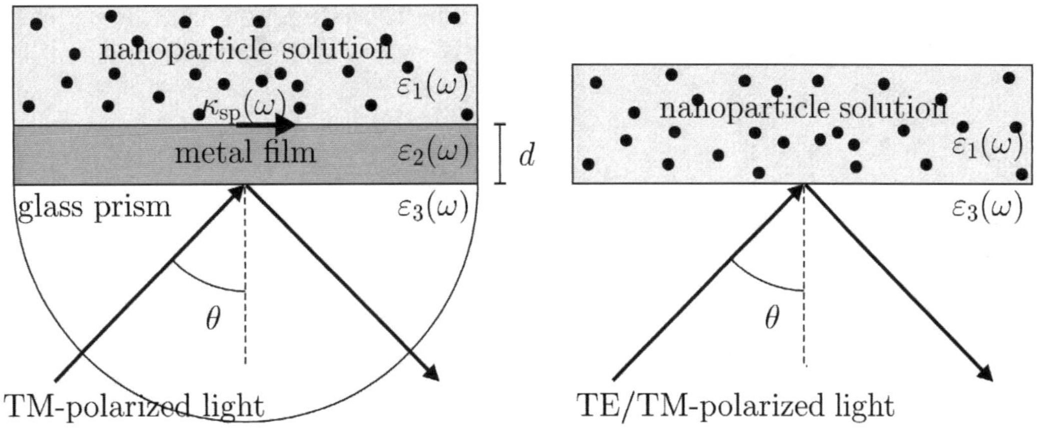

Figure 1. Kretschmann's configuration to study SPR resonance (left) and a set-up for direct reflectance measurement using two different planar polarizations (right).

We analyze the reflectance from arbitrary angle of incidence with two different linear polarizations using well-known Fresnel reflectance coefficients for air/colloid interface as shown in Fig. 1. The colloid is described with an effective dielectric function using the MG formalism. The remarkable feature with direct TE/TM polarized reflectance is that we can immediately obtain the complex effective dielectric function of the colloid after retrieval of the complex reflectivities for an arbitrary angle of incidence. We have shown before [12] that it is possible to obtain optical properties of nanoparticles in a liquid matrix in the SPR reflectance configuration as shown in Fig. 1 with the aid of the MEM for phase retrieval. However, the SPR experimental set-up has drawbacks such as contamination and oxidation of the metal film that supports plasmons. Therefore, the current set-up significantly simplifies the experiments as one can do a similar analysis from the reflectance without the SPR configuration as there is no need for the metal film.

The general problem setting in the phase retrieval problems is as follows: find a well-localized smooth complex-valued function f for a given modulus $|f|$. This is also the case for nanoparticle identification from liquid matrix by using reflectance data either in SPR or in a direct reflectance measurement mode. From reflectance measurement we get the amplitude of the reflectance but the phase information is lost. For a limited spectral range, which is the case here, we utilize the MEM, which originates from information theory [13]: entropy is a measure of the uncertainty represented by a probability distribution and the maximum entropy (ME) estimate is the least biased on the given information.

The phase retrieval procedure using the MEM includes fitting the measured reflectance spectrum into the maximum entropy model; for details of the method, see Ref. [7]. The unknown maximum entropy coefficients are obtained from a Toeplitz matrix with autocorrelation functions calculated from a discrete Fourier transform of the measured reflectance. This procedure always fits the maximum entropy phase function of the measured reflectance. The only feature of the actual phase function, which is not captured with the MEM, is the background phase information. This is the error phase function that is typically a much smoother function than the actual phase function of interest i.e. the actual phase function is a sum of the maximum entropy and error phase functions. The main features of the actual phase are always captured by the MEM. The key feature here is that we can use only the error phase function of the known host liquid similar to the SPR reflectance analysis [14].

Thus we can immediately obtain the effective complex dielectric function once the optical properties of the host liquid are known, i.e. no *a priori* knowledge is needed about the properties of the constituent nanoparticles.

Results and Discussion

In our simulations we will consider the direct reflectance geometry using two different linear polarizations. Figure 2 shows the complex effective dielectric functions of (a) dielectric and (b) metal nanoparticle colloids. An excellent agreement is observed between the exact values (solid lines) and ME estimates (dots). The resonance is well reproduced both for insulating and metallic nanoparticle colloids. It is worth noting that absorption is two orders of magnitude smaller in the case of dielectric nanoparticles than for metal nanoparticles. We can summarize our findings that we can always obtain the effective dielectric function of the colloid regardless of the actual colloid composition.

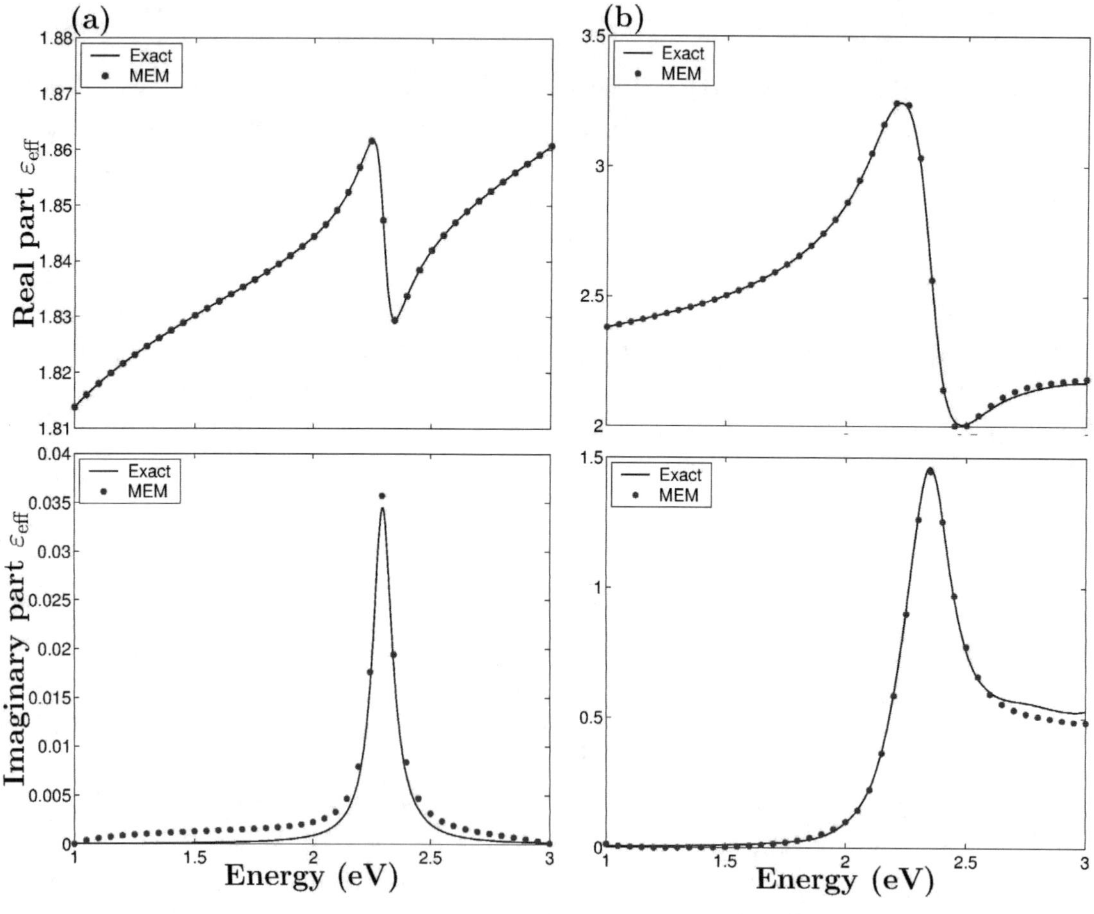

Figure 2. Retrieved effective complex dielectric functions for solution with (a) Lorentzian line-shape nanoparticles (parameters $\varepsilon_\infty = 2.5$, $\omega_0 = 2.3$ eV, $\omega^2_p = 0.1$ eV^2, and $\gamma = 0.1$ eV) and (b) gold nanoparticles with a volume fraction of 10% [14].

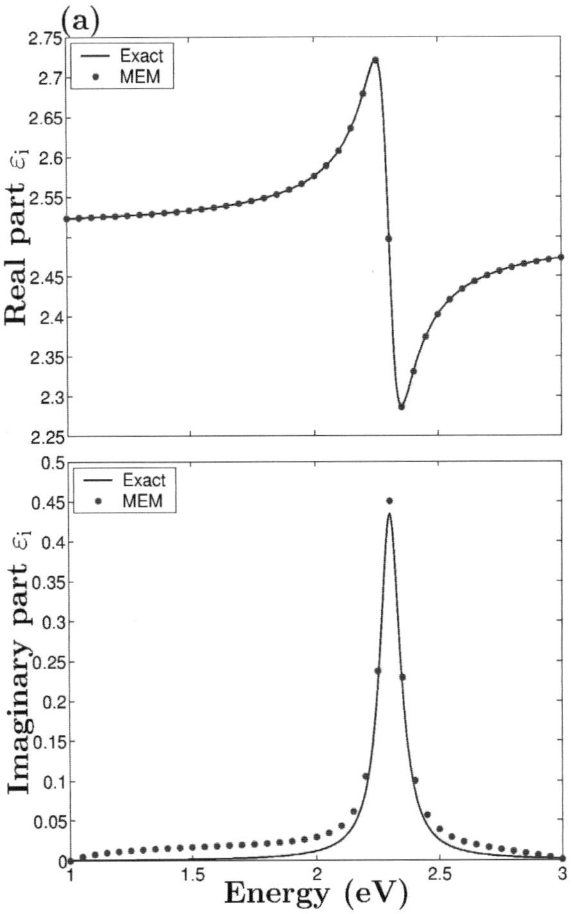

Figure 3. Retrieved complex dielectric functions for (a) Lorentzian line-shape nanoparticles (same parameters as in Fig. 2) and (b) gold nanoparticles [14].

Figure 3 displays the retrieved complex dielectric functions of the constituent nanoparticles calculated using the MG formalism. The dielectric nanoparticles in Fig. 3(a) are reproduced with an excellent accuracy especially in the vicinity of the resonance whereas the low frequency tail of metallic nanoparticles in Fig. 3(b) shows a significant divergence for the imaginary part from the bulk gold values. This divergence may be caused by numerical instability of the used MG model, which could be improved by using double precision throughout the phase retrieval calculations. However, the current numerical setup with single precision is closer to experiments, which inherently have limited accuracy.

We have investigated the source of the discrepancy especially in Fig. 3(b) and performed a similar analysis but using a double precision. The results for the effective and individual particle dielectric function are shown in Figs. 4 and 5 with 60° angle of incidence, respectively. We also present the errors of the retrieved values in Table 1: the corresponding error value with double precision is now reduced to 1.30×10^{-2} compared to a single precision value of 1.54×10^{-2}.

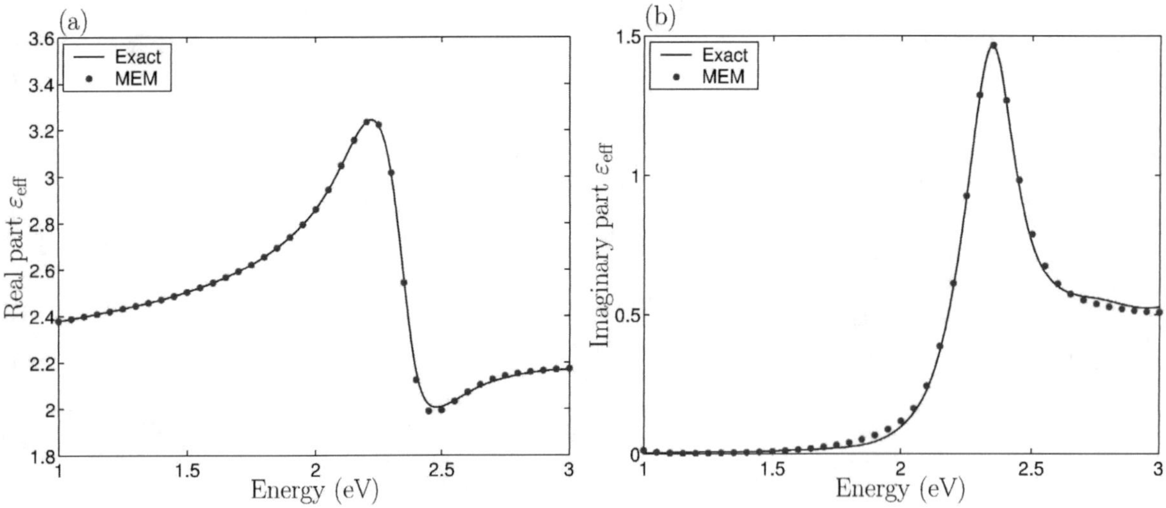

Figure 4. Effective dielectric function for gold colloid with double precision.

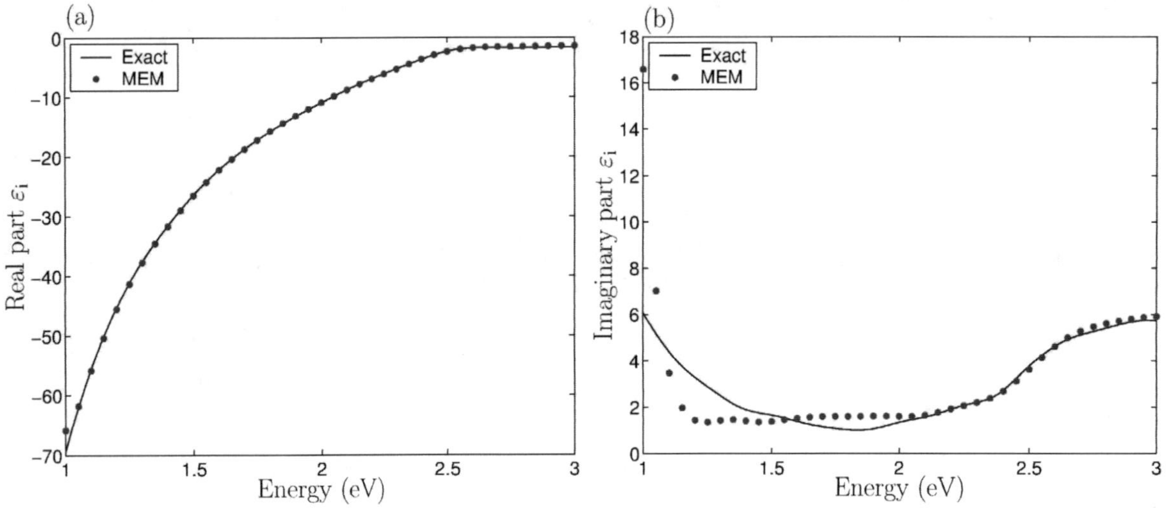

Figure 5. Retrieved dielectric function for gold nanoparticles with double precision.

Table 1. Average error of the MEM retrieved effective dielectric functions at 1–3 eV for different colloids with single and double precision (DP).

	15°	30°	45°	60°	75°
Dielectric ($\times 10^{-3}$)	1.16	1.16	1.20	0.993	1.05
Gold ($\times 10^{-2}$)	1.89	1.89	1.92	1.54	1.59
Gold DP ($\times 10^{-2}$)	–	–	–	1.30	–

Clearly we observe an improvement with double precision. However, the double precision does not completely solve the convergence issues for obtaining optical properties of individual metallic nanoparticles. Another factor is the used MG model to calculate the effective index of the colloid, which is then used backwards to obtain optical properties of individual nanoparticles. The MG formalism results in an equation with a small denominator value in the individual nanoparticle dielectric function equation. Hence, the retrieval of the constituent metal particle properties becomes vulnerable to small errors in the retrieved value for the effective index of a colloid. This is exactly seen in Fig. 3(b) as the small errors in the effective index are cumulated in the imaginary part of the particle dielectric function.

Conclusions

We have performed phase retrieval from TE/TM-polarized direct reflectance measurement both for dielectric and metallic nanoparticles. We observe an excellent agreement between the exact values and ME estimates with resonances well reproduced. It is worth emphasizing that we can always obtain the effective dielectric function of the colloid regardless of the actual colloid composition. Furthermore, we can also obtain the optical properties of individual nanoparticles if the colloid can be modeled with an effective medium theory. In the present case we use Maxwell Garnett model and obtain both dielectric and metallic nanoparticle dielectric functions with sufficient accuracy. The best accuracy is obtained with large angles of incidence but this is expected as reflectance becomes more sensitive to the optical properties of colloid above Brewster's angle of incidence.

As a conclusion we have shown that it is possible to obtain the complex dielectric function both for colloids and the individual dielectric or metallic nanoparticles from direct reflectance measurement using two different linear polarizations. The present method for direct reflectance measurement is experimentally simple and accurate for nanoparticle identification. The results here are applicable for any type of colloid independent of the actual colloid composition. We believe that the optical methods will find many applications in nanoparticle identification from water streams in the future.

Acknowledgments

This work has been funded by the Academy of Finland (grant no 250 122 & 256 263). J.J.S. wishes to thank the Japan Society for the Promotion of Science (JSPS) for an invitation to perform research at the Hokkai-Gakuen University in Sapporo, Japan in 2011–2012 under the 'FY2011 JSPS Invitation Fellowship Program for Research in Japan (Long-term)', Grant No. L-11537.

References

1. Maxwell Garnett, J. C. (1904), Phil. Trans. R. Soc. 203, 385.
2. Huang, X., Jain, P. K., El-Sayed, I. H. and El-Sayed, M. A. (2007), Nanomedicine 2, 681.
3. Weiss, J., Takhistov, P. and McClements, D. J. (2006), J. Food Sci. 71, R107.
4. Kretschmann, E. and Räther, H. (1971), Z. Naturforsch. A 23, 2135.
5. Liedberg, B., Nylander, C. and Lunström, I. (1983), Sens. Actuators 4, 299.
6. Räty, J., Peiponen, K.-E. and Asakura, T. (2004), *UV-Visible Reflection Spectroscopy of Liquids*, Springer: Heidelberg.
7. Lucarini, V., Saarinen, J. J., Peiponen, K.-E. and Vartiainen, E. M. (2005), *Kramers-Kronig Relations in Optical Materials Research,* Springer, Heidelberg.
8. Cornwell, T. J. and Evans, K. F. (1985), Astron. Astrophys. 143, 77.
9. Mannermaa, J. and Karras, M. (1989), Geophysica 25, 37.
10. Johnson, P. B. and Christy, R. W. (1972), Phys. Rev. B 6, 4370.

11. Partington, J. R. (1960), *An Advanced Treatise on Physical Chemistry*, vol. 4, Longmans, Green, New York.
12. Saarinen, J. J., Vartiainen, E. M. and Peiponen, K.-E. (2009), Sens. Actuators B: Chem. 138, 383.
13. Shannon, C. E. (1949), Bell System Technical J. 28, 656.
14. Saarinen, J. J., Uozumi, J., Vartiainen, E. M., and Peiponen, K.-E. (2012), Optics Letters 37, 2202.

(736d) Microfluidic Impedance Coulter Counter for Safety Applications

Joseph Parisi Jr.[1], Ying Wang[1], Christine Tartaglia[2] and Yu Lei[3], (1)Chemical, Materials, and Biomolecular Engineering, University of Connecticut, Storrs, CT, (2)University of Connecticut, (3)Chemical, Materials & Biomolecular Engineering, University of Connecticut, Storrs, CT

Microfluidic Impedance Coulter Counter for Safety Applications

Joseph Parisi, Ying Wang, Christine Tartaglia, Yu Lei

Department of Chemical, Materials, & Biomolecular Engineering, University of Connecticut, 191 Auditorium Road, Unit 3222, Storrs, Connecticut 06269

The need for highly sensitive, quick and reliable detection of pathogens in food and water sources is an important topic for the safety of our food and water supplies. Research in the area of microfluidics has targeted real time cell detection and has already solved some of these problems by creating various separation and detection devices. In this study we will develop an impedance based counting method for near real time detection of targets. Our device is focused on using a simple electrode design to measure resistance changes within our fluid with minimal post-processing of the collected data. The proof-of-concept will be demonstrated using a model particle and the analytical characteristics will be investigated. Its application for real pathogen detection will be explored.

(736e) Amperometric Detection and Quantification of Perchlorate in Groundwater Supplies Using a Highly Sensitive Nanostructured Electropolymerized PEDOT Biosensor

Ankush A. Gokhale, Department of Chemical Engineering & Materials Science, Michigan State University, East Lansing, MI, Jue Lu, Technova Corp, Lansing, MI and Ilsoon Lee, Department of Chemical Engineering and Materials Science, Michigan State University, East Lansing, MI

There has been a steady rise in the release of perchlorate (ClO_4^-), a prominent contaminant associated with the discharge of ammonium perchlorate from military and aerospace installations over the past few decades. Its remarkable recalcitrance to environmental degradation on account of its high activation energy coupled with high aqueous solubility has made it possible for ClO_4^- to persist in the ecosystem for a long time and make its way to the top of the food-chain. Severe cases of thyrotoxicosis are known to occur when the mammalian thyroid gland absorbs similar sized perchlorate ions instead of the iodide ions. Ever since the addition of ClO_4^- to the Contaminant Candidate list, there has been an increased effort to develop detection and quantification protocols for field tests that provide instant yet reliable results. In this study, we suggest the use of a portable, highly sensitive, fast acting enzymatic biosensor capable of measuring minute concentrations of perchlorate. Our perchlorate biosensor would consist of a sensing element in the form of perchlorate reductase, class of enzymes capable of reducing perchlorate to chlorite immobilized on a conductive polymer platform responsible for generating an amperometric response. Given their conductive nature, several polymers such as polyacetylene, polyaniline and polypyrrole have been used in the past for incorporation in sensor technology. In our work, we use poly(3,4-alkylenedioxypyrrole) (PEDOT), a conducting polymer resistant to nuclueophilic attack at the β-position. The controlled growth of PEDOT nanoarrays of desired morphologies using template assisted electropolymerization offers great potential to immobilize a host of bioactive components. In our work, we show the incorporation of perchlorate reductase on free-standing PEDOT nano-arrays using self-assembly techniques. The use of mediators in establishing faster electrical connection between the redox sites of the enzyme and the electrode surface is well-known. We propose to investigate the role of methyl viologen, a cationic mediator in shuttling electrons to establish fast yet accurate amperometric response. All electrochemical measurements were made using a CH instruments-model 650 A workstation using the conventional three electrode system.

(736f) Novel Sensor for Measuring Trace Impurities in Ultra Pure Hydrogen

Peter Bossard[1], **Jacob Mettes**[1], Fred Gornick[2] and Luis Breziner[3], (1)R&D, Power and Energy Inc, Ivyland, PA, (2)Chemistry, University of Maryland, New York, NY, (3)Power and Energy, Ivyland, PA

Parts per trillion analysis of ultrapure gases is currently done by Atmospheric Pressure Ionization Mass Spectrometry, APIMS, which requires sophisticated laboratory instrumentation and a specialized operator. When measuring contamination in hydrogen, a chemically pumped dense metal heated palladium membrane can concentrate the impurities, by eliminating most of the hydrogen gas while retaining all of the contaminants, leaving a remainder that can be analyzed with an RGA. To illustrate the method, a one part per trillion contaminant species in an accumulated 5 standard liter of ultrapure hydrogen sample that ends up in the closed 1 liter Residual Gas Analyzer chamber represents a partial pressure of 5*760*E-12 or 3.8E-9 Torr.

Associated manipulations and data interpretation are simple enough for the creation of an online sensor device highly desirable for gas quality monitoring in semiconductor manufacturing or in hydrogen filling stations for cars in the future.

Hydrogen removal with a palladium based membrane is gradient driven and sub-milliTorr hydrogen sample pressures can be achieved by chemical pumping downstream of the membrane. This is achieved by introducing air downstream of the membrane which oxygen irreversibly reacts with hydrogen to form predominantly OH ions. The concentrated 7.5 cc sample is introduced in a 1 liter RGA chamber, priory evacuated by a turbopump and then isolated by a valve causing a very reproducible pressure rise. Filament initiated chemistry and outgassing are slow processes under these low pressure conditions and can easily be distinguished from the actual contribution of the introduced sample by recording the initial step change in the contaminating gas' partial pressures. Trace contaminants in the UPH stream accumulate linearly with the sample time and are typically generated downstream of the hydrogen purifiers used in semiconductor manufacturing. Palladium purifier contamination mechanisms were identified and remedied by design and treatment of components downstream the purifier's membrane and are linked to the steel's carbon content and oxidation.

Ample experimental data will be provided demonstrating linearity of accumulation, calibration, chemistry in the RGA chamber and ppt level sensitivity for contaminating species.

(786a) Improving Detection Specificity of Label-Free Optical Biosensors

Ce Shi, Mork Family Department of Materials Science and Chemical Engineering, University of Southern California, Los Angeles, CA, Simin Mehrabani, Mork Department of Materials Science and Chemical Engineering, University of Southern California, Los Angeles, CA and Andrea M. Armani, Mork Family Department of Chemical Engineering and Materials Science & Ming Hsieh Department of Electrical Engineering-Electrophysics, University of Southern California, Los Angeles, CA

Improving detection specificity of label-free optical biosensors

Ce Shi[1], Simin Mehrabani[1] and Andrea M. Armani[1,2*]

[1]*Mork Family Department, University of Southern California,*

[2]*Ming Hsieh Department of Electrical Engineering-Electrophysics, University of Southern California,*

Los Angeles, California 90089, USA

**Corresponding author:armani@usc.edu*

Whispering gallery mode optical cavities fabricated on a silicon substrate from silica and silicon have demonstrated promise in the field of biodetection, both for fundamental biological investigations and as a diagnostic tool. For example, arrays silicon microring devices have performed detection and discrimination of multiple cancer markers in serum and silica resonators have been used to determine binding coefficients of proteins.[1-2] In both cases, the cavity surfaces were functionalized to detect the molecule of interest, and detection was performed by monitoring the change in the resonant wavelength of the cavity. Therefore, the specificity of detection relied on the accuracy of the functionalization molecule while the sensitivity was the result of the device performance or photon lifetime in the cavity. While there are numerous methods of achieving high performance optical cavities, often accurate discrimination between molecules is not possible, as multiple analytes can bind to the same targeting moiety, resulting in similar detection signals. Therefore, a complementary and simultaneous analysis method would further improve the reliability of this method.

This improvement can be achieved by analyzing both the resonant wavelength shift and the binding off-rates of the materials.[3] In the present work, toroidal optical microcavities are covalently functionalized with biotin, which binds robustly to streptavidin. Using solutions containing either free streptavidin or streptavidin functionalized polystyrene beads, a pair of experiments are performed to demonstrate this approach. First, free streptavidin in solution is detected and the off-rate is determined. In this control experiment, a single resonant wavelength shift is observed, as expected. Then a mixture of the two solutions is injected and detected. Surprisingly, a pair of shifts is detected. Based on mass transport calculations and a comparison with the initial control data, the first peak is the free streptavidin and the second peak is the polystyrene beads. Additionally, the two complexes have different dissociation constants, allowing further verification.

This work demonstrates the feasibility of combining resonant frequency shift with a kinetic and mass transport analysis to improve the reliability of resonant cavity-based detection.[3] This approach is very straightforward to implement in systems which involve fluid flow and could enable these devices to be used in solutions with high levels of contaminants.

(1) Washburn, A.; Gunn, L. C.; Bailey, R. C. *Analytical Chemistry* **2009**, *81*.

(2) Soteropulos, C.; Hunt, H.; Armani, A. M. *Applied Physics Letters* **2011**, *99*, 103703.

(3) Shi, C.; Mehrabani, S.; Armani, A. M. *Optics Letters* **2012**, *37*.

(786b) FRET and PET Combo Sensing in a Single Material: Expanding the Dynamic Range of an Ultra-Sensitive Nitroaromatic Explosives Assay

Ying Wang, Chemical, Materials, and Biomolecular Engineering, University of Connecticut, Storrs, CT and Yu Lei, Chemical, Materials & Biomolecular Engineering, University of Connecticut, Storrs, CT

Sensitive and selective detection of explosives is crucial for homeland security and civilian safety. In this research, a polyethylenimine polymer derivatized with pyrene moieties suitable for the fluorescence-based detection of nitroaromatic explosives (NAC) in aqueous systems is described. The system exhibits an exceptionally wide dynamic sensing range of 7 orders of magnitude (from 33 ppt to 225 ppm TNT or Tetryl). This broad range was achieved by the combination of two sensing mechanisms in a single material: a highly sensitive and selective FRET-based fluorescence quenching of the pyrene excimer band by a Meisenheimer complex formed by the reaction of the most electron-poor NACs with the amine-functionalized polymer. A less sensitive PET-based quenching of the fluorescence of monomeric pyrene moieties by the NACs covers the high concentration regime. The sensing material is suitable for a paper strip assay. Simplicity, selectivity, and the wide dynamic range suggest this material for explosives detection in the field.

(786c) The Controlling Parameters of Pyrene/Polymer Thin Films As Fluorescence Explosive Detecting Materials

Hyunsook Jang, Polymer Program, University of Connecticut, Storrs, CT, Ying Wang, Chemical, Materials, and Biomolecular Engineering, University of Connecticut, Storrs, CT, Yu Lei, Chemical, Materials & Biomolecular Engineering, University of Connecticut, Storrs, CT and **Mu-Ping Nieh**, Chemical, Materials & Biomolecular Engineering, Institute of Materials Science, University of Connecticut, Storrs, CT

The Controlling Parameters of Pyrene/Polymer Thin Films as Fluorescence Explosive Detecting Materials

A most recent report has shown that electrospun pyrene and polystyrene thin films results in high fluorescence and can be fast quenching (within 6 minutes) by nitro-aromatic explosive molecules. Moreover, the sensitivity (at the ppb level) and selectivity (specific to the nitro-aromatic compounds) are exceptional, making the materials an ideal candidate for explosive detection. Nevertheless, it has also been found that the preparation procedure of the thin films plays a key role in both fluorescence and quenching efficiency. In this presentation, we will discuss over a few parameters which effectively control the performance of the detection and provide some insight to the molecular interaction and aggregation structures of pyrene in the polymer thin films. The preliminary result indicates that polymer and evaporation rate of the solvent can strongly influence the fluorescence, suggesting that the high-performance structure obtained from the electrospun films is presumably dictated by the kinetic of the system.

(786d) Effects of Minor Asymmetries in Boundary Conditions of Cantilever Resonators On Sensing Properties

Blake N. Johnson and Raj Mutharasan, Department of Chemical and Biological Engineering, Drexel University, Philadelphia, PA

Engineering new cantilever sensors has pioneered various new sensing applications, including atomic-level microscopy and versatile bio-chemical sensing applications. In particular, dynamic-mode (resonating) cantilevers have become increasingly used since sensing principle involves direct transduction of added-mass by resonant frequency shifts. The canonical driving force for improving cantilever sensitivity suggests exploring designs that increase cantilever resonant frequency and reduce mass. Thus, various techniques have been examined, such as miniaturization, use of high-order resonant modes, and non-rectangular geometries. All such methods focus on the continuum of the structure (i.e. its size, motion, and geometry), but ignore the boundaries which present an equally rich potential for optimization. We demonstrate an alternative method to continuum engineering of such devices based on introducing minor perturbations in constrained cantilever boundaries (anchored regions) using finite element analysis and experimental measurements. The technique enhances cantilever resonant frequency, and thus, the realizable sensitivity of resonating cantilever sensors. The inclusion of asymmetric constrained boundaries also has a significant impact on the mode shapes and deflection magnitudes relative to those found in conventional cantilever sensors which contain ideal symmetric boundary conditions. Besides the potential for this phenomena as an applied technique for enhancing cantilever sensing properties, it is also important for describing experimental measurements in extremely small cantilevers fabricated near the precisions limits of micro-fabrication techniques where inherent asymmetry in the boundaries may arise. Therefore, understanding the effects of minor perturbations in constrained boundaries may become a central theme of future cantilever sensor engineering and design.

(786e) Electromagnetically Active Nanocomposite Metamaterial Biosensors

D. Keith Roper, Chemical Engineering, University of Arkansas, Fayetteville, AR, Phillip Blake, Department of Chemical Engineering, University of Arkansas, Fayetteville, AR and Drew Dejarnette, Microelectronics/Photonics Graduate Program, University of Arkansas

Title: Electromagnetically active nanocomposite metamaterial biosensors

Abstract Body: Novel active biosensor platforms offer both sensitive detection and real-time characterization of biological targets in genomics, proteomics, and biomedical theranostic applications. An example is quantized photon-exciton coupling which generates distance-dependent Forster resonance energy transfer in spectroscopic molecular rulers. We have recently distinguished quantized plasmon polarizabilities which induce extraordinary opto-electronic coupling with phase-dependent photon diffraction in ordered metalloceramic nanocomposite metamaterials. Metamaterials exhibit tunable electromagnetic functionality -- from simple iridescence in butterfly wings to radiofrequency cloaking -- due to coherent interference from multidimensional structuring of suitable condensed-matter composites. Sensitivity enhancement from tunable electromagnetic coupling in metamaterials appears more than 10-fold relative to sensitivities in conventional planar or nanoparticle sensors based on surface plasmon resonance. Our lab has hybridized 'top-down' lithography and 'bottom-up' metallization approaches to fabricate nanocomposite metamaterial biosensor platforms. Using scanning electron microscopy and microspectroscopy we show that prototype devices based on these platforms exhibit opto-electronic sensitivities consistent with model predictions. These advances in modeling, fabrication, and prototype analysis are important milestones toward developing electromagnetically active nanocomposite metamaterials as biosensor platforms for next-generation disease diagnosis and personalized medicine.

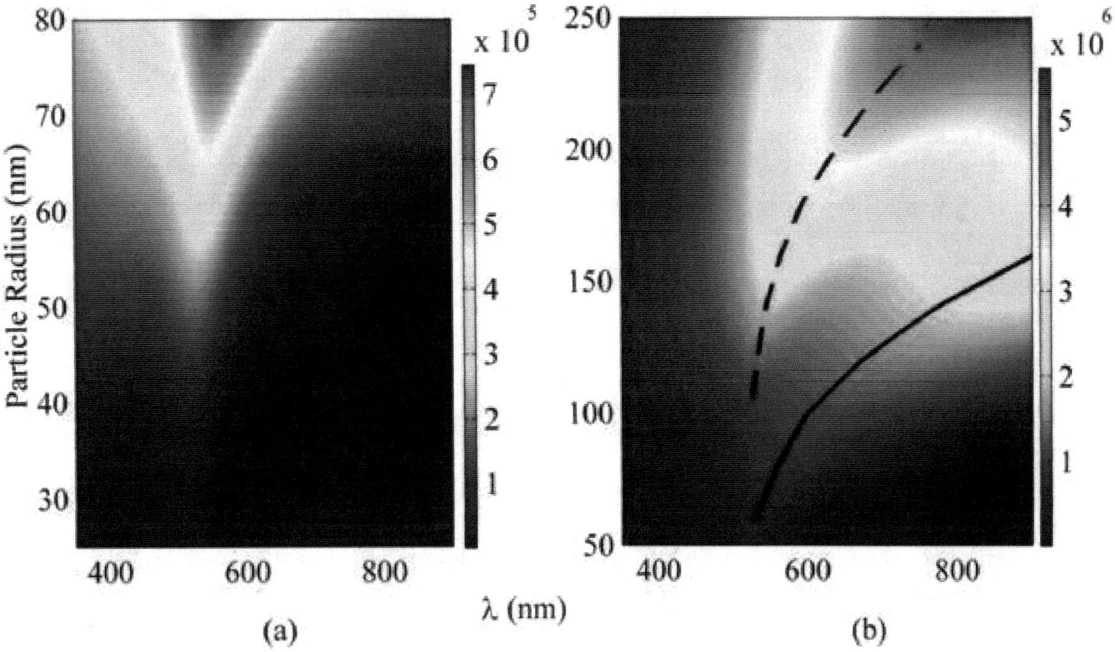

(786f) Nanoporous Membrane Molecular Sensors Based On Ion Currents

Hsueh-Chia Chang[1], Satyajyoti Senapati[1], Sunny Shah[2], Li-Jing Cheng[3] and Zdenek Slouka[3], (1)Chemical & Biomolecular Engineering, University of Notre Dame, Notre Dame, IN, (2)Chemical & Biomolecular Engineering, University of Notre Dame, (3)Chemical and Biomolecular Engineering, University of Notre Dame, Notre Dame, IN

We report a new molecular (nucleic acid and protein) microfluidic sensor platform that is economical to mass produce, label-free, multi-target (>5), rapid (<10 min), selective (SNP) and sensitive (pM)----and is hence most suitable for portable point-of-care diagnostic devices. This new sensing platform employs fabricated or synthesized nanoporous membranes in PDMS, glass or polymer chips. The synthesis techniques include solgel crystallization and photocuring. When a DC field is applied across a nanoporous ion-selective membrane (or a fabricated nano-channel or nanostructure assembly), a molecular-sized charge polarized layer develops on the depletion side after concentration polarization. The presence of target molecules in this polarized layer can hence sensitively change the limiting and overlimiting currents of the membrane. The overlimiting current shift is particularly large (~1 V) because large nucleic acids can change the hydrodynamic slip length and hence suppress the surface electrokinetic vortices responsible for the overlimiting current. Oppositely charged small molecules, on the other hand, can produce a bipolar membrane like a PN semiconductor junction that can produce a large water-splitting ion current signal when a high field develops at the carrier-depleted junction with a reverse bias. Without using electron transfer to detect the molecules, we eliminate the need for SAM and prevent any robustness issue due to spurious reactions, as in classical electrochemical sensing platforms. This new molecular sensing technology also reduces the transport-limited assay time by several orders of magnitude because the high field in the polarized layer can attract large nucleic acids by dielectrophoresis and the depletion region can concentrate smaller ones in a flowing stream at the membrane sensor location. We also use flow to linearize hairpin structures/collapsed globules and remove non-specifically bound molecules. An optimum shear rate is found empirically as a function of the target molecule size and the probe-target association constant-----and is accurately captured by an activated theory for hybridization. Since each membrane is only a micron-wide strip, extension to more than 10 targets can be easily done. We have successfully tested and quantified this platform in double blind tests for RNAs of 4 Dengue virus seroltypes, microRNA of ovarian cancer cells, cytokine and interferon-gamma signal proteins, Enterococcus, E coli, Brucella, Listeria, MRSA, Pseudomonas, Meningitis and four aquatic invasive species: quagga mussel, D bugenesis, E sinensis, L fortunei. RNA detection is particularly exciting because of microRNA as a new cancer biomarker and the fact that many bacteria have tens of thousands of mRNA but only one copy of DNA implies that PCR amplification is not necessary even with only one pathogen per sample. Prototype devices based on this platform is being developed and commercialized. We also describe current extension of this sensor platform to massively parallel target detection and regenerable real-time sensors for immune cell cultures and remote applications.

Yossifon, G. and Chang, H.-C., “Selection of Nonequilibrium Overlimiting Currents: Universal Depletion Layer Formation Dynamics and Vortex Instability”, Phys Rev Lett., 101, 254501(2008).
Chang, H.-C. and Yossifon, G., "Understanding Electrokineitcs at the Nanoscale--a Perspective", Biomicrofluidics, 3, 012001 (2009).
Basuray, S. , Senapati, S., Ajian, A. , Mahon, A. R. and Chang, H.-C., “Shear and AC Field Enhanced Carbon Nanotube Impedance Assay for Rapid, Sensitive and Mismatch-Discriminating DNA Hybridization”, ACS Nano, 3, 1823 (2009).
Cheng, I.-F., S. Senapati, X. Cheng, S. Basuray, H-C Chang and H.-C. Chang, “A Rapid Field-Use Assay for Mismatch Number and Location of Hybridized DNAs”, Lab-on-a-Chip,10, 828-831 (2010).
Mahon, A. R., Barnes, M. A., Senapati, S., Feder, J., Chang, H.-C. and Lodge, D. M., “Molecular Detection of Invasive Species in Heterogeneous Mixtures using a Carbon Nanotube Platform”, PLOS One, 6, 17280 (2011).
Senapati, S., Basuray, S., Slouka, Z., Cheng, L.-J. and Chang, H.-C., "A Nanomembrane Based Nucleic Acid Sensing Platform for Portable Diagnostics" , Topics in Current Chemistry, 304, , 153-169(2011).
Chang, H.-C., Yossifon, G. and Demekhin, E. A. , “Nanoscale Electrokinetics and Microvortices: How

Microhydrodynamics Affects Nanofluidic Ion Flux", Annual Rev of Fluid Mech, 44, 401-426 (2012).

Cheng, L.-J. and Chang, H.-C., "Microscale pH Actuation by Splitting Water", Biomicrofluidics, 5, 046502 (2011).

(786g) Love-Mode Surface Acoustic Wave Devices Fabricated Using ZnO:V Thin Films As Waveguiding Layers On 90°-Rotated ST, X-Cut Quartz Substrates for Sensing Applications

Mandek B. Richardson[1], Devajyoti Mukherjee[2], Pritish Mukherjee[2], Sarath Witanachchi[3] and Venkat R. Bhethanabotla[4], (1)Chemical & Biomedical Engineering, University of South Florida, Tampa, FL, (2)Physics, University of South Florida, (3)Physics, Unversity of South Florida, (4)Chemical Engineering, University of South Florida, Tampa, FL

Love-mode surface acoustic wave devices fabricated using ZnO:V thin films as waveguiding layers on 90 rotated ST-X-cut quartz substrates for sensing applications

Mandek Richardson[a], Devajyoti Mukherjee[b], Pritish Mukherjee[b], Sarath Witanachchi[b], and Venkat R. Bhethanabotla[a]

[a] *Sensors Research Laboratory, Department of Chemical & Biomedical Engineering, University of South Florida, Tampa, Florida 33620, USA*

[b] *Center for Integrated Functional Materials, Department of Physics, University of South Florida, Tampa, Florida 33620, USA*

Love-mode surface acoustic wave devices (SAW) have been successfully used as gas and liquid sensors because of their high sensitivity [1, 2]. A Love-wave device consists of a thin waveguiding layer deposited on a SAW substrate that supports shear-horizontal (SH) particle polarization. For the Love-mode to exist the shear acoustic velocity of the waveguiding layer must be less than the shear acoustic velocity of the substrate [3]. Some common materials implemented as guiding layers are SiO_2 [4] and polymethyl-methacrylate (PMMA) [5]. In recent years, ZnO thin films have proven to be a suitable guiding layer by providing higher sensitivity than SiO_2 [6]. In this work, 90 rotated ST-X-cut quartz SAW substrates were chosen because of their low temperature coefficient of delay (TCD) when combined with ZnO films at room temperature, high phase velocity (5060 m/s) and the existence of surface skimming bulk waves (SSBWs) [7, 8]. Particle polarization of SSBWs is parallel to the surface and transverse to the direction of wave propagation. This allows the SSBW mode to be used in liquid sensing applications because the absence of particles displaced normal to the surface prevents the radiation of energy into the adjacent liquid [9]. Deposition of a layer, such as ZnO, onto the surface transforms the SSBW to a surface wave, thereby increasing sensitivity by confining the acoustic energy within several wavelengths from the surface.

We chose ZnO as a waveguiding layer because of its low acoustic velocity (2747 m/s) compared to ST-X quartz [10], piezoelectricity, low acoustic absorption, and high chemical, mechanical and thermal stability. Deposition of ZnO thin films was done by pulsed laser deposition (PLD); this technique yields high quality thin films with oriented crystallites at relatively low substrate temperatures of about 200 °C to 400 °C. There are several reports that have investigated the effects doping has on the properties of both undoped and doped ZnO thin films used in Love-wave sensors [11-15]. Dopants studied include magnesium (Mg), calcium (Ca), strontium (Sr), lithium (Li), and barium (Ba). Theoretical and experimental analyses have shown that the highest sensitivity occurs at an optimum ratio of the waveguide thickness to operating wavelength [6]. Additional increases in sensitivity occur for a certain dopant level because adding the dopant increases the surface roughness. The positive correlation between surface roughness and sensitivity is caused by several mechanisms: generation of non-laminar motion in the liquid, trapping of liquid by cavities or pores, and conversion of in-plane surface motion to surface normal motion in the liquid [12]. In this work we decided to study the effects of using vanadium (V) as a dopant into ZnO as compared to undoped ZnO thin films. Dilute doping of V into ZnO thin films enhances their dielectric and piezoelectric properties which would lead to higher electromechanical coupling (K^2) in ZnO based Love-SAW devices [16, 17]. Therefore, a SAW sensor with a ZnO film doped with the optimum concentration of V will have increased performance due to better crystallinity and stronger piezoelectric properties. Furthermore, the roughness of the film can also be increased by changing the deposition parameters to also improve sensor sensitivity [17].

In this work we present a Love-wave device based on V- doped ZnO films on 90 rotated ST-X quartz substrates. ZnO films were deposited using PLD with a thickness range from 0.5 to 2.0 μm. The films were characterized by XRD, SEM, and energy dispersive X-ray spectroscopy (EDS) to determine the optimal deposition conditions. SAW sensors containing ZnO films doped with different concentrations of V were compared to SAW substrates containing un-doped ZnO films at the same thicknesses. Experiments included measuring sensor insertion loss and sensitivity to mass loading to determine the effectiveness of doping ZnO films with V.

[1] D. W. Branch and S. M. Brozik, "Low-level detection of a Bacillus anthracis simulant using Love-wave biosensors on 36 degrees YX LiTaO3," *Biosensors & Bioelectronics,* vol. 19, pp. 849-859, Mar 15 2004.

[2] C. Zimmermann, D. Rebière, C. Déjous, J. Pistré, E. Chastaing, and R. Planade, "A love-wave gas sensor coated with functionalized polysiloxane for sensing organophosphorus compounds," *Sensors and Actuators B: Chemical,* vol. 76, pp. 86-94, 2001.

[3] E. Gizeli, N. J. Goddard, C. R. Lowe, and A. C. Stevenson, "A Love Plate Biosensor Utilizing a Polymer Layer," *Sensors and Actuators B-Chemical,* vol. 6, pp. 131-137, Jan 1992.

[4] G. Kovacs, G. Lubking, M. Vellekoop, and A. Venema, "Love waves for (bio)-chemical sensing in liquids," *IEEE 1992 Ultrasonics Symposium Proceedings,* pp. 281-285, 1992.

[5] G. L. Harding and J. Du, "Design and properties of quartz-based Love wave acoustic sensors incorporating silicon dioxide and PMMA guiding layers," *Smart Materials & Structures,* vol. 6, pp. 716-720, Dec 1997.

[6] R.-C. chang, S.-Y. Chu, C.-S. Hong, and Y.-T. Chuang, "A study of Love wave devices in ZnO/Quartz and ZnO/LiTaO3 structures," *Thin solid films,* vol. 498, 2006.

[7] K. H. Yen, K. L. Wang, and R. S. Kagiwada, "Efficient bulk wave excitation on ST quartz," *Electronics Letters,* vol. 13, pp. 37-38, 1977.

[8] T. I. Browning and M. F. Lewis, "A New Class of Quartz Crystal Oscillator Controlled by Surface-Skimming Bulk Waves," presented at the 31st Annual Symposium on Frequency Control, 1977.

[9] K. Kalantar-Zadeh, D. A. Powell, W. Wlodarski, S. Ippolito, and K. Galatsis, "Comparison of layered based SAW sensors," *Sensors and Actuators B-Chemical,* vol. 91, pp. 303-308, Jun 1 2003.

[10] S.-Y. Chu, W. Water, and J.-T. Liaw, "An investigation of the dependence of ZnO film on the sensitivity of Love mode sensor in ZnO/quartz structure," *Ultrasonics,* vol. 41, pp. 133-139, 2003.

[11] R.-C. Chang, S.-Y. Chu, P.-W. Yeh, C.-S. Hong, P.-C. Kao, and Y.-J. Huang, "The influence of Mg doped ZnO thin films on the properties of Love wave sensors," *Sensors and Actuators B: Chemical,* vol. 132, pp. 290-295, 2008.

[12] W. Water and Y.-S. Yang, "The influence of calcium doped ZnO films on Love wave sensor characteristics," *Sensors and Actuators A: Physical,* vol. 127, pp. 360-365, 2006.

[13] P. Wu, J. Zhong, N. Emanetoglu, Y. Chen, S. Muthukumar, and Y. Lu, "Li diffusion in epitaxial (11<img src="/fulltext-image.asp?format=htmlnonpaginated&src=FP6Q016PRN3N378W_html\11664_2004_Article_52_TeX2GIFIE1.gif" border="0" alt=" $$\bar 2$$ " />0) ZnO thin films," *Journal of Electronic Materials,* vol. 33, pp. 596-599, 2004.

[14] W. Water, T.-H. Fang, L.-W. Ji, T.-H. Meen, and Y.-S. Yan, "Surface Morphology and Liquid Sensor Sensitivity of Barium-Doped ZnO Thin Film," *Journal of Science and Innovation,* vol. 1, pp. 25-32, 2011.

[15] W. Water, S.-F. Wang, Y.-P. Chen, and J.-C. Pu, "Calcium and strontium doped ZnO films for Love wave sensor applications," *Integrated Ferroelectrics: An International Journal,* pp. 37-41, 2005.

[16] Y. C. Yang, C. Song, X. H. Wang, F. Zeng, and F. Pan, "Giant piezoelectric d33 coefficient in ferroelectric vanadium doped ZnO films," *Applied Physics Letters,* vol. 92, 2008.

[17] T. Dhakal, D. Mukherjee, R. Hyde, H. Srikanth, P. Mukherjee, and S. Witanachchi, "Enhancement in Ferroelectricity in V-Doped ZnO Thin Film Grown Using Laser Ablation," *Materials Research Society Symposium Proceedings,* vol. 1199E, 2010.

(786h) Delay Path Modification of a Surface Acoustic Wave Sensor On ST-X Quartz: A Computational Study and Comparison to Experiments

Mandek B. Richardson, Chemical & Biomedical Engineering, University of South Florida, Tampa, FL, Subramanian Sankaranarayanan, Center for Nanoscale Materials, Argonne National Lab, Argonne, IL and Venkat R. Bhethanabotla, Department of Chemical Engineering, University of South Florida, Tampa, FL

Delay path modification of a surface acoustic wave sensor on ST-X quartz: A computational study and comparison to experiments

Mandek Richardson[a], Subramanian K.R.S. Sankaranarayanan[b], and Venkat R. Bhethanabotla[a]

[a] *Sensors Research Laboratory, Department of Chemical & Biomedical Engineering, University of South Florida, Tampa, Florida 33620, USA*

[b] *Center for Nanoscale Materials, Argonne National Laboratory, Argonne, Illinois 60439, USA*

Surface Acoustic Wave (SAW) devices can be used as direct, label free biosensors that monitor the interaction between a receptor and its target in real time, through changes in the properties of the traveling wave (i.e., frequency, phase, and amplitude) ADDIN EN.CITE *Branch2007534[1]53453417Darren W. BranchThayne L. EdwardsIEEE Ultrasonics SymposiumIEEE Ultrasonics Symposium260-2632602007*[1]. An ideal biosensor will be selective to the species of interest, robust, have high sensitivity, and consume a small amount of power. When it comes to sensing clinically relevant concentrations devices with high sensitivities are paramount. In addition, due to an increase in remote sensing applications, minimizing power consumption is also a key design consideration.

SAW sensors having insertion loss as high as 20-30 dB are common. Due to its low electromechanical coupling coefficient, K^2, this is usually the case when quartz is the substrate. ST-X quartz generates surface skimming bulk waves (SSBWs) ADDIN EN.CITE *Yen197713[2, 3]131317Yen, K. H.Wang, K. L.Kagiwada, R. S.Electronics LettersElectronics LettersElectron Lett37-38132quartzultrasonic transducersST quartzbulk wave excitationtransducer configurations19770013-519410.1049/el:19770027Browning197714141447T. I. BrowningM. F. Lewis31st Annual Symposium on Frequency Control258-2651977*[2, 3] when the direction of propagation is rotated 90° to its x-axis, and these SSBWs penetrate into the bulk material, but never reach the bottom surface, as they travel along the delay path between the IDTs. Particle polarization of SSBWs is parallel to the surface and transverse to the direction of wave propagation which allows this wave mode to be used in liquid sensing applications since the lack of normal particle displacement doesn't radiate energy into the adjacent liquid ADDIN EN.CITE *Kalantar-Zadeh20034[4]4417Kalantar-Zadeh, K.Powell, D. A.Wlodarski, W.Ippolito, S.Galatsis, K.Kalantar-Zadeh, K
RMIT Univ, Sch Elect & Comp Engn, City Campus,GPO Box 2476V, Melbourne, Vic 3001, Australia
RMIT Univ, Sch Elect & Comp Engn, City Campus,GPO Box 2476V, Melbourne, Vic 3001, Australia
RMIT Univ, Sch Elect & Comp Engn, Melbourne, Vic 3001, AustraliaSensors and Actuators B-ChemicalSensor Actuat B-ChemSensors and Actuators B-ChemicalSensor Actuat B-ChemSensors and Actuators B-ChemicalSensor Actuat B-Chem303-308911-3layered surface acoustic wave (saw)wave guiding layermass sensitivityelectromechanical coupling coefficientwave acoustic sensorsloveliquidssurface2003Jun 10925-4005ISI:000183039200049<Go to ISI>://000183039200049Doi 10.1016/S0925-4005(03)00091-1English*[4]. ST-X quartz was chosen for our study because of its applicability in liquid biosensing ADDIN EN.CITE *Browning197714[3]141447T. I. BrowningM. F. Lewis31st Annual Symposium on Frequency Control258-2651977*[3], it's cost, and it can be readily etched by deep reactive ion etching (DRIE) ADDIN EN.CITE *Chen2008630[5]63063017H. ChenC. FuJournal of Micromechanics and MicroengineeringJournal of Micromechanics and Microengineering18101050012008*[5].

To overcome high insertion loss and/or increase sensitivity to mass loading several approaches have been utilized: the incorporation of reflective gratings ADDIN EN.CITE *Morgan199116[6]16166D. P. Morgan1991New YorkElsevier*[6], grooves and corrugated surfaces ADDIN EN.CITE *Auld19799[7]9910Auld, B. A.Yeh, B. H.1979 Ultrasonics Symposium1979 Ultrasonics Symposium786-7901979197910.1109/ultsym.1979.197312*[7], the application of waveguide materials ADDIN EN.CITE *Gizeli199210[8]101017Electra GizeliAdrian C. StevensonNicholas J. GoddardChristopher R. LoweIEEE Transactions on Ultrasonics, Ferroelectrics, and Frequency ControlIEEE Transactions on Ultrasonics, Ferroelectrics, and Frequency Control657-6593956571992*[8], and optimization of transducer designs that support unidirectional wave propagation ADDIN EN.CITE *Yamanouchi198415[9]151517K. YamanouchiH. FuruyashikiUltrasonic Symposium ProceedingsUltrasonic Symposium Proceedings68-713681984*[9]. Based on those methods Cular et al. reported on a novel device that contained an array of square microcavities etched into the delay path between SAW IDTs ADDIN EN.CITE *Cular20086[10]6617Cular, S.Sankaranarayanan, S. K. R. S.Bhethanabotla, V. R.Cular, S
Univ S Florida, Dept Chem & Biochem Engn, Sensors Res Lab, Tampa, FL 33620 USA
Univ S Florida, Dept Chem & Biochem Engn, Sensors Res Lab, Tampa, FL 33620 USA
Univ S Florida, Dept Chem & Biochem Engn, Sensors Res Lab, Tampa, FL 33620 USAApplied Physics LettersAppl Phys LettApplied Physics LettersAppl Phys LettApplied Physics LettersAppl Phys Lett9224love wave2008Jun 160003-6951ISI:000256934900126<Go to ISI>://000256934900126Artn 244104
Doi 10.1063/1.2949553English*[10]. Simulations performed on SAW devices modified with microcavities show that it is a more effective way to decrease insertion loss and increase sensor sensitivity by creating larger coherent reflections of the generated SSBW and reducing conversions of the surface wave into bulk modes [10, 11].

In this study we compare simulations and experiments performed on SAW sensors having an array of microcavities with varying dimensions contained in the delay path. The best case microcavity device was then compared with conventional SAW and Love-wave sensors. We show that fabricated SAW devices with microcavities in the delay path confirm simulations, lower insertion loss, and increase sensitivity to mass loading. Furthermore, our results indicate that SAW devices with microcavities can be used as a sensor that is highly sensitive to the detection of biomolecules and has decreased power consumption to be used in clinical diagnostic systems.

Figure 1: Schematic of a SAW delay line with an array of microcavities lying in the path between the input and output IDTs

References

ADDIN EN.REFLIST [1] D. W. Branch and T. L. Edwards, "Love Wave Acoustic Array Biosensor Platform for Autonomous Detection " *IEEE Ultrasonics Symposium,* pp. 260-263, 2007.

[2] K. H. Yen, K. L. Wang, and R. S. Kagiwada, "Efficient bulk wave excitation on ST quartz," *Electronics Letters,* vol. 13, pp. 37-38, 1977.

[3] T. I. Browning and M. F. Lewis, "A New Class of Quartz Crystal Oscillator Controlled by Surface-Skimming Bulk Waves," presented at the 31st Annual Symposium on Frequency Control, 1977.

[4] K. Kalantar-Zadeh, D. A. Powell, W. Wlodarski, S. Ippolito, and K. Galatsis, "Comparison of layered based SAW sensors," *Sensors and Actuators B-Chemical,* vol. 91, pp. 303-308, Jun 1 2003.

[5] H. Chen and C. Fu, "An investigation into the characteristics of deep reactive ion etching of quartz using SU-8 as a mask," *Journal of Micromechanics and Microengineering,* vol. 18, 2008.

[6] D. P. Morgan, *Surface-Wave Devices for Signal Processing*. New York: Elsevier, 1991.

[7] B. A. Auld and B. H. Yeh, "Theory of Surface Skimming SH Wave Guidance by a Corrugated Surface," in *1979 Ultrasonics Symposium*, 1979, pp. 786-790.

[8] E. Gizeli, A. C. Stevenson, N. J. Goddard, and C. R. Lowe, "A Novel Love-Plate Acoustic Sensor Utilizing Polymer Overlayers," *IEEE Transactions on Ultrasonics, Ferroelectrics, and Frequency Control,* vol. 39, pp. 657-659, 1992.

[9] K. Yamanouchi and H. Furuyashiki, "Low-Loss SAW Filter Using Internal Reflection Types of New Single-Phase Unidirectional Transducer," *Ultrasonic Symposium Proceedings,* vol. 3, pp. 68-71, 1984.

[10] S. Cular, S. K. R. S. Sankaranarayanan, and V. R. Bhethanabotla, "Enhancing effects of microcavities on shear-horizontal surface acoustic wave sensors: A finite element simulation study," *Applied Physics Letters,* vol. 92, Jun 16 2008.

[11] R. Singh, S. K. R. S. Sankaranarayanan, and V. R. Bhethanabotla, "Enhanced surface acoustic wave biosensor performance via delay path modifications in mutually interacting multidirectional transducer configuration: A computational study," *Applied Physics Letters,* vol. 95, Jul 20 2009.

AUTHOR INDEX